Making Sense of Human Anatomy and Physiology

Making Sense of Human Anatomy and Physiology

A Learner-Friendly Approach

Earle Abrahamson
and Jane Langston

Chichester, England

North Atlantic Books
Berkeley, California

First published in 2017 by
Lotus Publishing
Apple Tree Cottage, Inlands Road, Nutbourne, Chichester, PO18 8RJ
North Atlantic Books
Berkeley, California

Anatomical Drawings Amanda Williams
Photographs all taken by Jane and Mark Langston unless otherwise indicated
Text and Cover Design Wendy Craig
Printed and Bound in Malaysia by Tien Wah Press

Making Sense of Human Anatomy and Physiology: A Learner-Friendly Approach is sponsored and published by the Society for the Study of Native Arts and Sciences (dba North Atlantic Books), an educational nonprofit based in Berkeley, California, that collaborates with partners to develop cross-cultural perspectives, nurture holistic views of art, science, the humanities, and healing, and seed personal and global transformation by publishing work on the relationship of body, spirit, and nature.

North Atlantic Books' publications are available through most bookstores. For further information, visit our website at www.northatlanticbooks.com or call 800-733-3000.

Disclaimer
Every effort has been made to include the most accurate and up-to-date information in this publication. However, the authors would be grateful for any errors to be brought to their attention. Neither the authors nor the publishers can take responsibility for misuse of this information or for injury caused by inappropriately applied treatment. Please consult a healthcare professional before applying any of the methods discussed in this text.

British Library Cataloguing-in-Publication Data
A CIP record for this book is available from the British Library
ISBN 978 1 905367 70 2 (Lotus Publishing)
ISBN 978 1 623171 73 5 (North Atlantic Books)

Library of Congress Cataloging-in-Publication Data
Names: Abrahamson, Earle, author. | Langston, Jane, author.
Title: Making sense of human anatomy and physiology : a learner-friendly approach / Earle Abrahamson and Jane Langston.
Description: Berkeley, California : North Atlantic Books, [2017]
Identifiers: LCCN 2016056329 (print) | LCCN 2017005251 (ebook) | ISBN 9781623171735 (paperback) | ISBN 9781623171742 (ebook)
Subjects: LCSH: Human physiology. | Human anatomy. | BISAC: MEDICAL / Anatomy. | SCIENCE / Life Sciences / Human Anatomy & Physiology. | HEALTH & FITNESS / Reference.
Classification: LCC QP34.5 .A25 2017 (print) | LCC QP34.5 (ebook) | DDC 612--dc23
LC record available at https://lccn.loc.gov/2016056329

Contents

Foreword

As I flew to London, I was nervous to present a new anatomy seminar, *Immaculate Dissection (ID)*, to a group of eager physiotherapists, trainers, chiropractors and fitness enthusiasts. At ID, we had created a seminar that filled in gaps in anatomy education for movement professionals. Our students took a chance on a new method of anatomy instruction, with body painting, palpation and movement assessments. Sitting front-row center for our London premiere was Jane Langston, the co-author of *Making Sense*.

Throughout the course, Jane eagerly asked questions about the body paintings and lecture materials, and she was quick to volunteer for the corrective exercise portions. I naturally gravitated towards her to ask about her training. She informed me of her goals in helping improve education of anatomy in every discipline that studies it.

As Jane described this book you are about to read, I knew she had created a necessary game changer in the way anatomy is taught. This book is not just for the student but for the anatomy educator.

This topic hits home for me. As a gross anatomy instructor with five colleges, I see the foundational anatomy getting lost as students progress into becoming clinicians.

Something is missing in anatomy instruction.

Making Sense covers this ground, by combining several different learning methodologies into one concise program. Students are applying the anatomy – not just memorizing where it is in space. *Making Sense* also closes learning gaps by providing multiple learning styles. This helps the educator and the student improve efficiency in the anatomy instruction.

If you are already studying anatomy and physiology, you'll find *Making Sense* helps you to solidify your understanding with its innovative approach to learning.

If this is your first experience learning anatomy and physiology, you potentially will have much frustration saved by the use of *Making Sense*.

And if you are an anatomy instructor, you already know the importance of an anatomy text. Perhaps you have room to grow by using *Making Sense* as one of your primary references. This book closes the gaps between text and the carryover of anatomy into professional life.

Kathy Dooley, DC, MSc
Creator, Immaculate Dissection
Instructor, Albert Einstein College of Medicine

About this Book

Rationale and Notes on How to Use the Text

> *Education is the kindling of a flame, not the filling of a vessel.*
>
> Socrates

The paradox ... There is more known, in every discipline, than students in any course can learn!

As a result, in most courses students are asked to learn too much, with the seemingly paradoxical result that too little is learned and retained.

If students can't learn it all, what should they learn? This text outlines the concepts of the big ideas. Big ideas are the key themes necessary to master the foundation knowledge. Through the compilation of learning methodologies, the text provides guidance on how to navigate difficult learning situations and learning barriers. This provides a revolutionary portal into the functional learning and application of anatomy from a learner-centred perspective.

Anatomy and physiology are a wonderful exploration and study of the human body that allow for an understanding of the processes of living and those that sustain life. There are multiple books and resources that explain the systematic approach to learning about the human body across levels of learning and learners. Few texts examine the core dynamics of learning and teaching anatomy and physiology.

This book engages the reader by presenting anatomy and physiology as a lived experience, one that requires understanding of terms and concepts. The book presents anatomy as a language and uses the principles of language learning to clarify meaning in anatomy and physiology study. In a learner-friendly approach, the chapters of the book take the reader through a journey of both studying and making sense of the content in anatomy and physiology. The book provides a road map for understanding problems and issues in studying the subject matter, provides useful insight into practical and effective assessment techniques, explores the subject matter from a learning-approach perspective, and considers different methods of teaching to understand how best to convey the message and meaning of anatomy and physiology. Like all books on teaching and learning, the emphasis of this text is on the quality process of delivering anatomy and physiology so that learners learn not only to pass an examination or assessment, but more importantly retain the fundamental building blocks of anatomical study and application. The final chapters provide useful resources that can be used to further explore concepts, assessment, learning activities and applications.

This book evolved from a desire to publish a text that truly aided the learning process. The authors, both of whom have vast experience in teaching and learning anatomical science, were concerned that few texts fully acknowledged the troubled spaces and complexities inherent in the study of anatomy and physiology. Most texts tend to present colourful artistic drawings of the body systems, but, in reality, what are the key learning objectives that students need to master in order to appreciate the art behind the science? Placing themselves in the student view, the authors have revisited their experiences of learning and provide insight and understanding into new and innovative, often practical, ways of kindling the flame for learning anatomy and physiology in practice. Through reflective practice and careful thought about the application of anatomy and physiology in a multitude of environments, the content of this book was born. The chapters mirror the experiential journeys taken by the authors in developing their knowledge and toolkits. It is through critical analysis, challenge and collaboration that true learning takes place.

Anatomy and physiology study is about learning to settle into the unsettled, and learning to ask new questions about the way our bodies work and fail. Through the chapters of the book, learners will hopefully become impassioned and empowered to take responsibility for their learning and make active choices in how they learn, so they, in effect, become their own teachers and learning facilitators.

Use this book to discover the love for learning, the challenge to develop new and innovative learning and teaching methods, and the courage to experiment with ideas and subject matter so that different configurations within learning can emerge. In the words from the *Sound of Music*'s 'Doe-Re-Mi,' 'When you know the notes to sing, you can sing most anything.' This book, essentially, is about learning to orchestrate the notes so that you can learn to produce different melodies.

Throughout this text we have purposely used the words 'student' and 'learner' interchangeably to illustrate a dichotomy between levels of knowledge development and mastery. We position the student as learner and engage the learner with studentship. The chapters are carefully aligned to allow for repetition and emphasis of key learning points, strategies and activities that are signposted within other chapters of the text.

The text is organised into two distinct themes. The first introduces the learning of anatomy and physiology and focuses on multiple dimensions for learning, enabling students and teachers to make sense of learning dynamics. The second considers challenges in learning, and directs learners towards assessment and revision strategies for anatomy and physiology. These themes are interconnected and flow throughout the text. The book concludes with a resource chapter identifying key learning and teaching activities and websites that enrich the learning experience for both student and teacher. Some chapters are written expressly for the teacher, whilst others contain elements applicable to both learner and teacher. In reading or studying anatomical images, one may be inclined to zoom in on the details and lose sight of the bigger picture. Anatomy and physiology are often best taught and learned by understanding the bigger picture before we become lost in or inspired by the finer details.

We hope this text will become a portal for learning and a critical friend for reflecting on learning.

Acknowledgements

Writing a book is not only about positioning words on a page: it requires time, effort and space to think through ideas, capture experiences and navigate the many territories that surround the subject matter. The writing experience is not devoid of difficulties, which often demand additional time, energy, creativity and collaboration. To the writer the journey could appear arduous and lonely, not really knowing when or possibly how it will end. It is in these particular situations that the writer's support network plays an important role. For me personally, my support provided me the opportunity to create the space and time to plan and develop the content and chapters for this text.

My co-author, Jane, enabled me to consider multiple lenses and acknowledge the many layers and sublayers that make anatomy and physiology exciting to learn and teach. To my wife Emma, who constantly challenged my beliefs and encouraged me to seek out debate, develop arguments and reconfigure what it is I truly believe: I thank you dearly. To my wonderful children Benjamin and Oliver: thank you for giving me the time to write, the freedom to think, the inspiration to create, and your love to envelop my passion for the learning and teaching of anatomy and physiology. I hope this book will inspire you to fulfil your dreams and life's ambitions. To my brother Michael, thank you for teaching me to appreciate the lessons that life teaches you. To my academic community, scholars, practitioners, professional organisations, teachers and students: thank you for providing a framework for me to deconstruct theory, apply knowledge and experience, and create a text that demonstrates the process for learning and teaching. To Jon and his team at Lotus Publishing, thank you for your invaluable assistance in bringing our text to life.

The creation of this text allowed me to value the scholarship of teaching and learning, and work towards not only improving practices but, more importantly, transforming them. The value of this text lies in the ability for communities of practice and learning to disseminate the messages and practices within the chapters.

I hope you enjoy the learning and knowledge transfer and use the text to encourage others to do the same. Finally I dedicate this book to the memory of my late parents, Josephine and Charles Abrahamson.

Earle Abrahamson

I dedicate this book in memory of my parents, John and Elizabeth Davies, who showed me that a strong work ethic, good-humoured persistence and perseverance gets the job done.

I am a self-confessed 'studyoholic'; my love for anatomy and physiology is only matched by my passion for learning. I was fortunate to have an inspirational teacher who spent endless hours teaching us how to learn Latin, rather than just throwing a new language at a bunch of twelve-year-olds. So, Mr Melville, you have probably no idea the impact you made on my future, and for that, I thank you.

I learned my passion in physiology and pathology as a teenager, working in the pathology departments at St Thomas' Hospital, London, and to the staff there I give grateful thanks that you sowed the seed of information and gave it a nurturing environment that allowed it to grow and blossom into the secure knowledge I possess today.

My support network of husband, son and colleagues has given me the encouragement to write, technical wizardry in mending my laptop and a sounding board for ideas, so to Mark and David Langston, Gina Lilley and Taz Faruqi: I couldn't wish for better family and friends.

My co-author, Earle, saw something in me I didn't recognise for myself, until he coerced me into writing with him; thank you Earle for your vision, ideas and trust.

And finally, it is with gratitude that I consider the impact that all my students and learners at The Amatsu Training School have made on me, and how by humbly teaching, I learn.

Jane Langston

1 Laying the Foundations

The body is intricately simple and simply intricate.

Dr George Goodheart

The art of learning anatomy and physiology is similar to that of building a house. How solid is the ground on which we are building? How deep should the foundations be? How many layers of bricks should we build in order for the room to be tall enough for our use? Just as the architect and the engineer work with the builder to construct detailed plans that must be followed in order to build a structure that resembles the initial design sketched by the architect, so must the anatomy and physiology course follow carefully thought-out plans. Depending on the amount of information needed for the end result, just as the house must be built on solid ground, the foundations of knowledge must be sufficient to withstand and support the correct layers of knowledge. Just as the house has all its services of electricity, gas, water and sewerage, in addition to its bricks and mortar, so must the anatomy and physiology course have all the required body systems in place, such as the nervous, respiratory, lymphatic, renal, digestive and musculoskeletal systems.

Each building that is constructed is handed over to its future occupants with a technical manual explaining the hardware of the building. This can be likened to the study of anatomy, with its intricate detail of how each organ or tissue is constructed. Occupants are also given an operations manual, explaining what to do when things go wrong. This is analogous to the study of physiology and pathology, with their explanations of how the body works and what happens when it goes wrong. And just as an apprentice in the building trade chooses a specialism to become a plumber, electrician, bricklayer or carpenter, anatomy and physiology students will go on to use the information learned on their course to become physiotherapists, sports therapists, doctors, vets, nurses, massage therapists and paramedics, to name but a few.

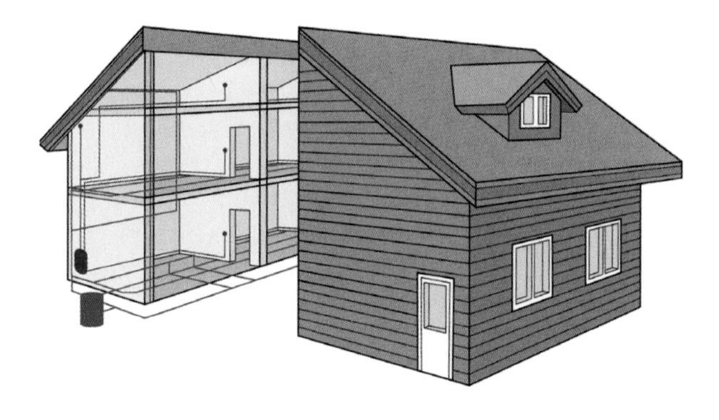

Figure 1.1: The systems of the body can be compared to those of a house.

My builder rang me to give me an update on delivery dates for some materials for the extension he was building for me. After only ten seconds, it dawned on me that I simply didn't speak 'building'. What on earth was a lintel? Something to boil up and turn into soup? Soffits? What were soffits? A new brand of fabric conditioner? And what was Celotex? Surely we didn't need any explosives to make our living room larger? By the time I had unpicked the language my builder rattled off to me in his broad Suffolk accent, I realised I actually couldn't remember a single thing he had said as all of these items had names that were completely foreign to me. This is exactly what a new anatomy student may encounter when confronted with terms such as 'Greater Trochanter'. Is that a village in Scotland? And as for Islets of Langerhans, surely they are a holiday destination in the West Indies! Tell them the greater trochanter is the big knobbly bit found at the top of the leg, and they will absolutely understand what and where it is, and can expand their knowledge by thinking about its attached muscles. And explain that the Islets of Langerhans are the insulin-producing cells of the pancreas that don't work properly in diabetes, and things begin to make sense. Likewise, my builder had to tell me he was ordering the concrete bits for the tops of my windows and the white boards that are behind my gutters, and that he would be buying packs of insulation to put between the roof beams. If he had said that in the first place, we would have had a much shorter conversation, and I would have been able to visualise the items on the delivery van.

Later in this book we will introduce the need to understand the language of anatomy and the need to learn and understand its taxonomy. In order to match teaching and learning to the understanding, we must use language-appropriate textbooks. Use starter-level books for beginners to anatomy and physiology rather than heavy duty tomes that go into too much detail too soon in the learning process. Just as a language is learned gradually, through conversation and the written word, so anatomy and physiology become familiar and friendlier as the correct terms are used appropriately. Find textbooks that describe anatomical structures in simple and understandable terms – this isn't dumbing down, it is simply setting a clear pathway to learning.

Aim to introduce a different textbook for each level of learning, with more detailed anatomical language used by the higher-level textbooks. This encourages the learners to use different resources and allows their anatomical language skills and comprehension to grow at the appropriate rate. It also offers the students a variety of learning styles, and stops them from getting stuck.

Teachers of anatomy and physiology must make sure they use 'differentiated language' when teaching. When mentioning an anatomical structure that has a long, complicated and unfamiliar name, say its technical term, and, in the same breath, use a more familiar and softer descriptive. Eventually, the students become familiar with the technical term, and less reliant on the more everyday version.

Each individual system is taught step by step to allow the details of the system to be understood and learned. However, each system is interdependent and interconnected with others.

In the schematic diagram in Figure 1.3, adjacent body systems have links and parallels with each other, yet each is its own system, which has its own functions.

Example:

Insertion of the psoas muscle can be described as attaching to the inside of the top of the femur, which is called the lesser trochanter. A trochanter is a knobbly piece on a bone that acts as a muscle attachment point.

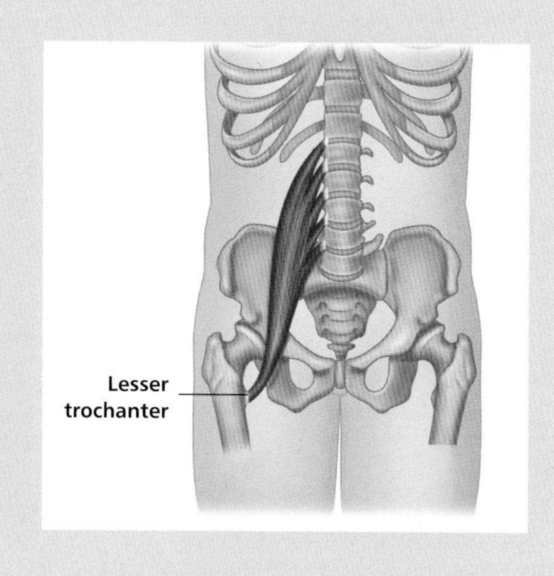

Figure 1.2: The psoas.

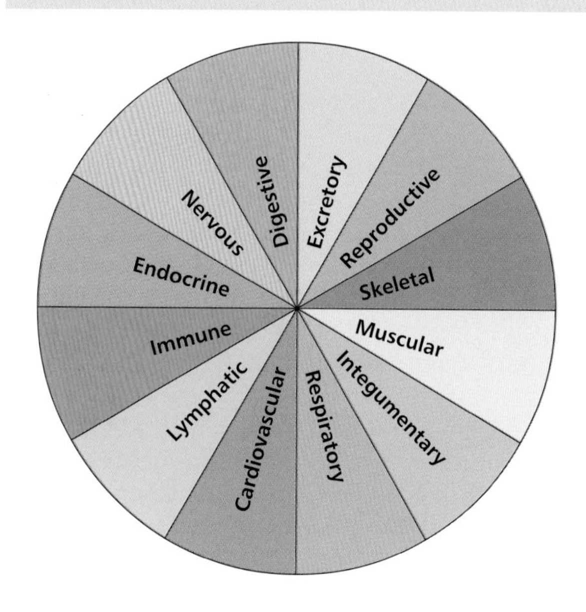

Figure 1.3: Laying the foundations, system by system.

The functions of the body systems can be learned by thinking of them as parts of a house. A radiator, pump or piece of copper piping is pointless on its own, but together they form the central heating system of the house.

Musculoskeletal Systems

The most obvious structural parts of a house are the bricks and mortar, and their accompanying joists, rafters and foundations. These could represent the musculoskeletal systems, which give support and structure to the body, and provide a framework to which the other organs and systems attach.

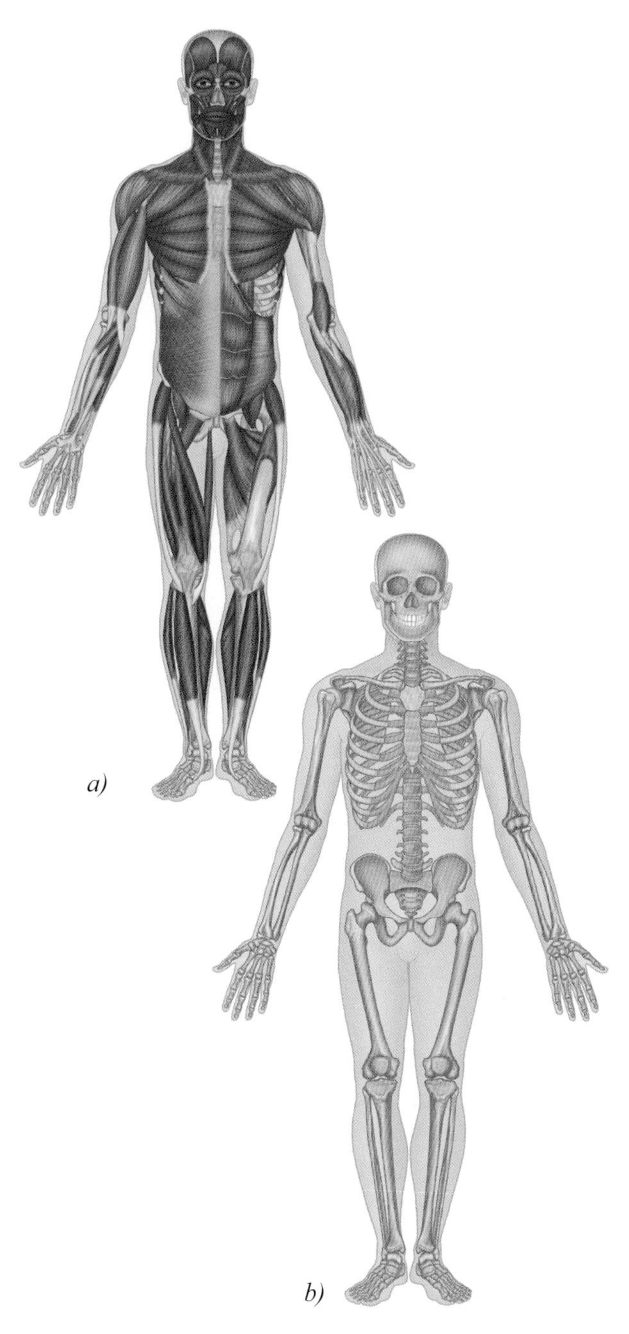

Figure 1.4: a) Muscular system, b) skeletal system, c) rafters and blocks represent bone, creating the house shape, with ties holding the rafters together, acting as ligaments, and, d) plasterboard nailed to timber-framework at strategic points, much like muscle attaching to bone.

The internal load-bearing walls, timber-frames, metal struts and rafters gives the solid structure to the house, and these represent our skeletal system. Rafters and blocks construct the overall shape and will represent bone. These give attachment points for plasterboard, ties and external walls. Ligaments, the reinforced straps for our joints, could be portrayed as the truss connector plates which are used to hold the rafters and joists in place.

Internally, muscles join to bones at strategic points called attachments, so could be likened to plasterboard nailed to the joists and timber-framework. Using this analogy does not truly demonstrate the ability of the bones to move when the muscles are contracted. However, we can consider how settlement cracks occur, when the house's internal structure moves and settles, causing the plaster and plasterboard to crack and move.

Mortar and cement represent the connective tissue, which forms a continuum around the body, touching and supporting the cells that make up the bones and muscles.

Integumentary System (Skin)

The outside of the house may be coated in tiles, paint, pebble-dash or stucco to protect the bricks and mortar, and may be painted in different colours. This represents the *integumentary system* or *skin* – the outer coating of the house representing the epidermis, and the layers of insulation in the cavity walls being likened to the fatty layers of the dermis, through which cables and pipework may be flowing.

The paintwork of a house protects it from any water ingress, and stops creepy crawlies coming in through the bricks. In the same way, the skin's function is to give a protective layer, to stop the body drying out or getting soggy, and to protect it from bacteria or viruses that may land on the skin. It acts as insulation for the body, and provides a method of using sunlight to convert various chemicals to vitamin D. Although the house doesn't do this, solar panels may be on the outside of a building, using solar energy to produce electricity. Both the outside of the house and the inside are decorated with paint. This is synonymous with the epithelium and the endothelium that continue throughout the body.

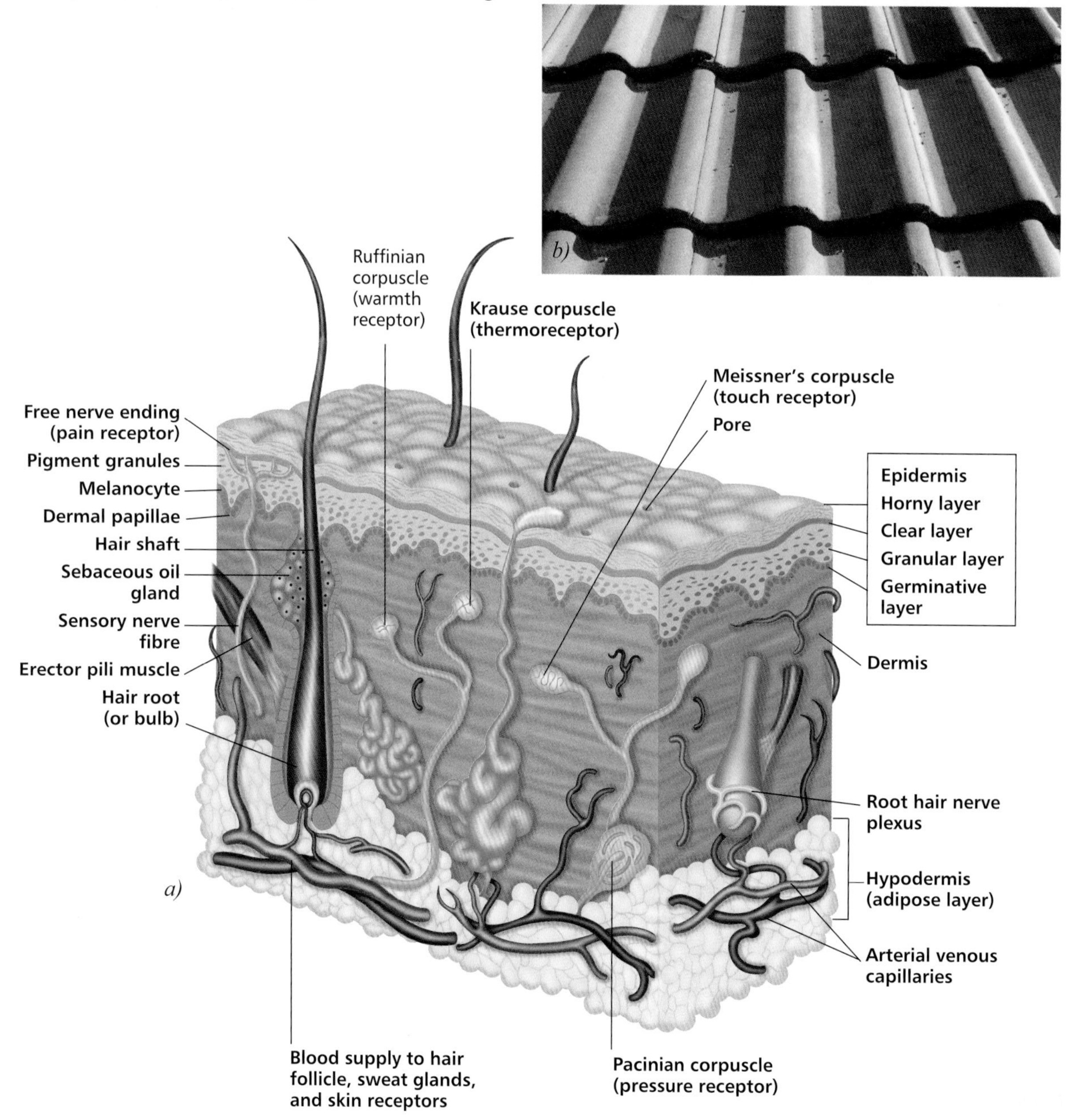

Figure 1.5: a) Integumentary system, b) roof tiles represent the outer layer of the house, or skin.

Respiratory System

All good housing needs ventilation, so that fresh air is drawn into the building, and contaminated or spent air is expelled. The air bricks act as the nose and mouth, allowing air into the building, and the exhaust vents of the central heating systems and extractor fans send steamy, unwanted air back outside. These represent the *respiratory system*, which brings air from outside the body, via the nose and mouth, into the lungs, where gaseous exchange takes place, oxygen is taken into the body and moistened, and carbon-dioxide-rich air is expelled. If the ventilation of a building is insufficient then toxic gases may build up, which can be fatal – and this can be seen in respiration too.

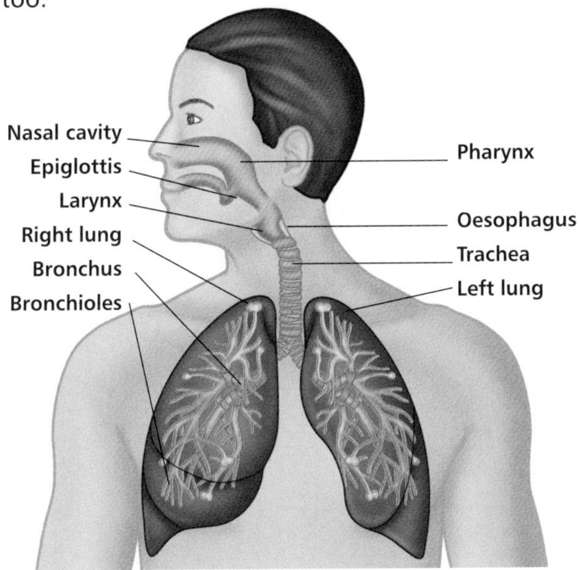

Figure 1.6: Respiratory system.

Nervous System

The house is powered by electricity, and this is supplied around the property by cabling, junction boxes and switches, and these are protected by fuses. In our bodies, the electrical system is analogous to the *nervous system*, with its neurons and synapses. Electrical cabling has very thick insulation, similar to the myelin sheaths surrounding some neurons. If this myelin disappears owing to disease then the message does not flow through the neuron, just as when the copper wire is exposed when it shouldn't be, the electrical pathway short-circuits and blows a fuse. If there is a power cut in a house, the equipment within it grinds to a halt and everything stops; this happens in the body too. If the nervous system stops, all bodily functions halt, as the control of them is affected and thus they are unable to continue functioning. As technology advances, more and more the electricity in our homes will be controlled by computers, just as our bodies are controlled by the brain.

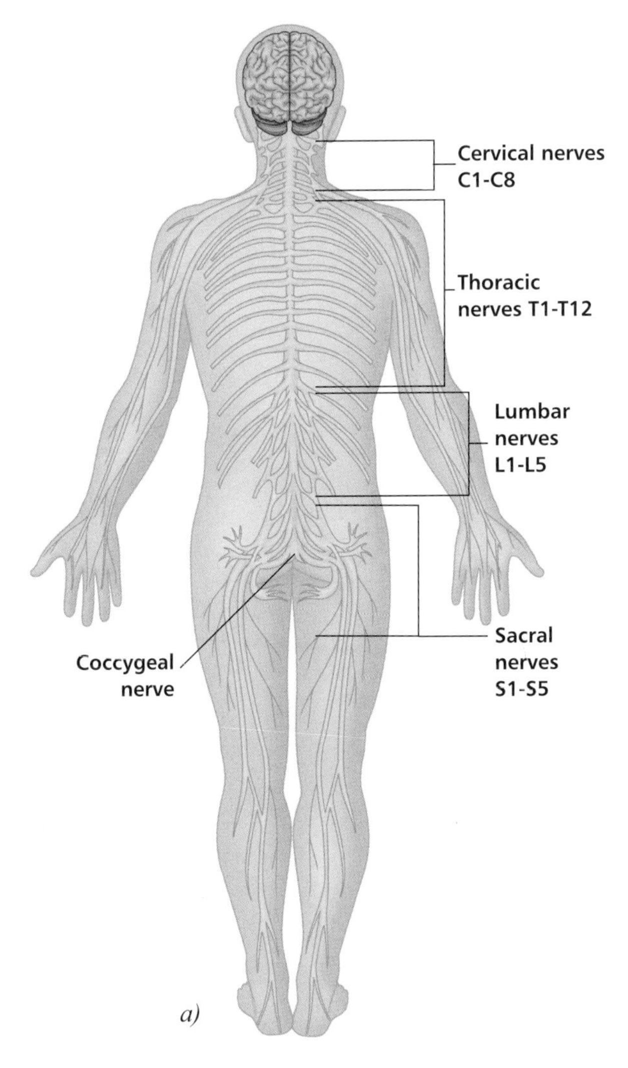

Figure 1.7: a) Nervous system, b) wires in the electrical system controlling the workings of a house represent the nerve fibres, with the plastic coating of each wire acting as the myelin sheath.

Endocrine System

We need to keep the temperature of our home nice and stable, so we don't freeze or melt. We also need to make sure that the gas levels in the house are kept just right so that the heating system fires and we don't get poisoned by toxic fumes. We need to have the right amount of cleaning agents and chemicals too, so that germs don't grow in the toilets, limescale doesn't build up in the sinks and weeds don't grow in the garden. In our body, we need to keep both our temperature and our chemical environment stable, and this stability is called homeostasis.

Homeostasis is defined as the ability to maintain internal stability in an organism to compensate for environmental changes. The body maintains this by having neurological sensors that detect change, just as the house has carbon monoxide and smoke detectors which alert to dangerous levels of gas or smoke. Our house has thermostats which control the temperature of the heating, and our bodies have the hypothalamus, a neuroendocrine gland that, amongst other things, controls the level of heat in the body. The *endocrine system* comprises endocrine tissue/glands that secrete chemical messages called hormones, which maintain the levels of other chemicals and functions in the body, a bit like the water softeners and cleaning agents that maintain the house.

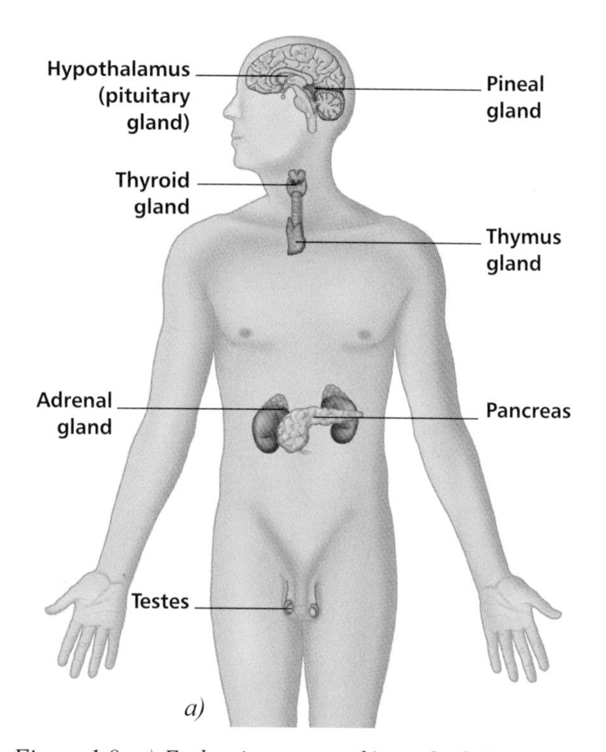

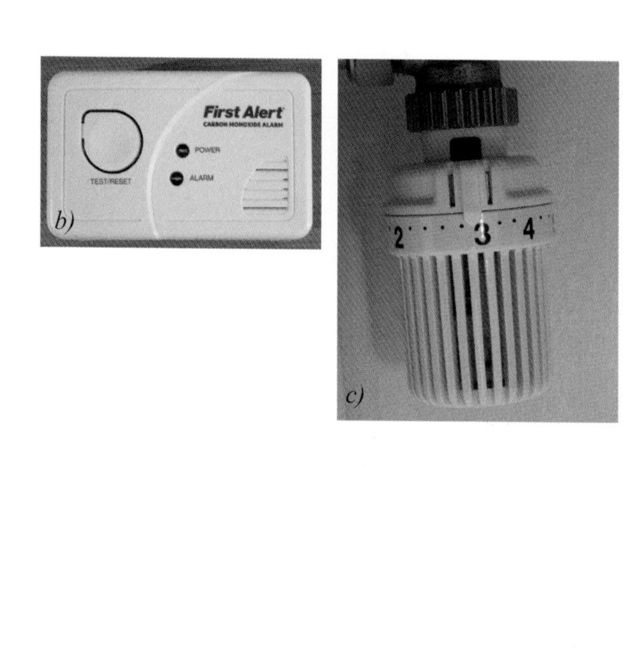

Figure 1.8: a) Endocrine system, b) our body have neurological sensors that detect change, much like a carbon monoxide detector alerts the home to dangerous levels of gas, c) thermostats control the heating of the house, much like the hypothalamus in the brain controls the level of heat in the body.

Circulatory, Lymphatic and Immune Systems

The passage of fluids throughout a house relies on the plumbing and its pumps, and the water and waste fluids which flow through it. Similarly, the *circulatory system* of our body contains pipework in the form of arteries, veins and capillaries, and it is pumped via the heart. If any of these pipes springs a leak, we bleed, just as our heating and water system leaks when pipes burst. We even say that we 'bleed' a radiator to get any air out of it.

The fluids in our house are of two types – clean water and soiled water. The passage of drinking water would represent the passage of blood through the arteries, and into our capillaries and interstitial fluid.

Figure 1.9: Fluids.

Soiled water would represent a part of the circulatory system that is also part of the *lymphatic system*; giving pipework to drain the body of its interstitial fluid. The lymphatic system can be seen as part of the *immune system*; the spleen and lymph nodes offer a suitable location for lymphocytes and other cells of the immune system.

The immune system is also the body's burglar alarm system, checking for entry of bacteria and viruses, and setting off alarms in the form of immunoglobulins or complement cascades when it detects any bugs.

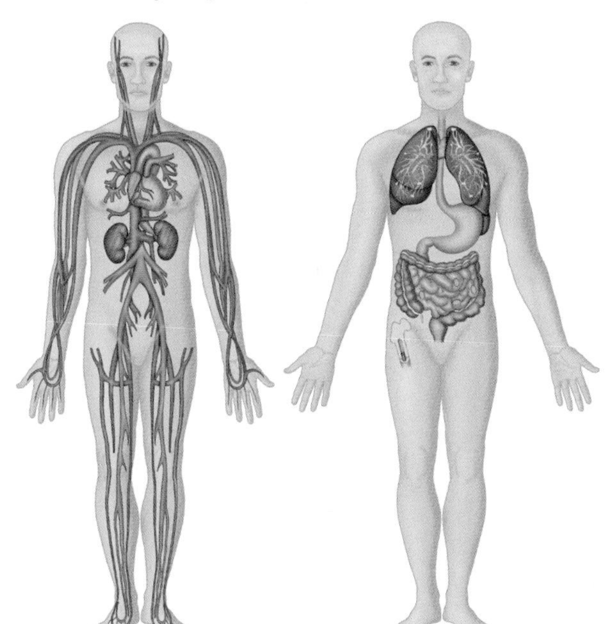

Figure 1.10: a) Circulatory system, b) immune system.

Digestive System

The centre of the home is the kitchen, where people gather to prepare food and perhaps to eat too. Our kitchen cupboards store the food, ready to be chopped and prepared so it can be eaten. It is seasoned and cooked so that it is made more palatable and is easily digested. The waste-disposal unit is connected to the sewerage system, so that any unwanted or unused food is sent out of the house. Likewise, the *digestive system* of the body chomps food using the mouth, teeth and tongue; makes it more digestible using enzymes in the stomach and duodenum; and absorbs it into the body via the small intestine. And the excess and unwanted residue is turned into faeces and expelled via our own sewerage system, the large intestine.

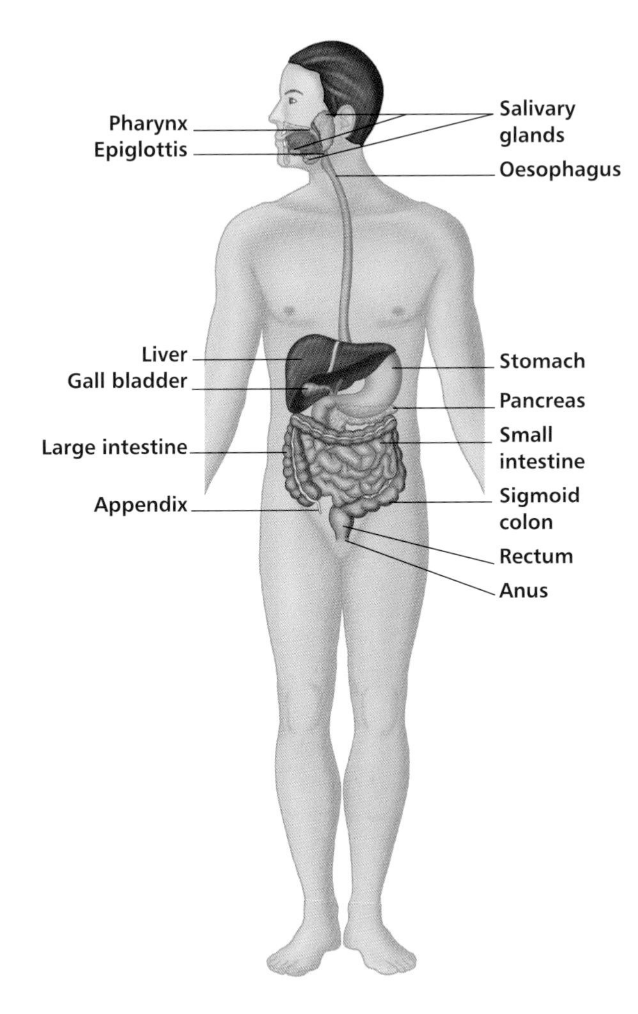

Figure 1.11: Digestive system.

Excretory System

As vast quantities of 'junk' in our house get thrown into the bin or skip, so in our bodies, depending on the type of waste we need to excrete, we have designated organs to deal with the waste: sweat glands in our skin, and the colon and rectum. Liquid waste in the house is expelled via the drains and toilet, just as our body filters our waste fluids and expels them via the kidneys and bladder of the *excretory system*.

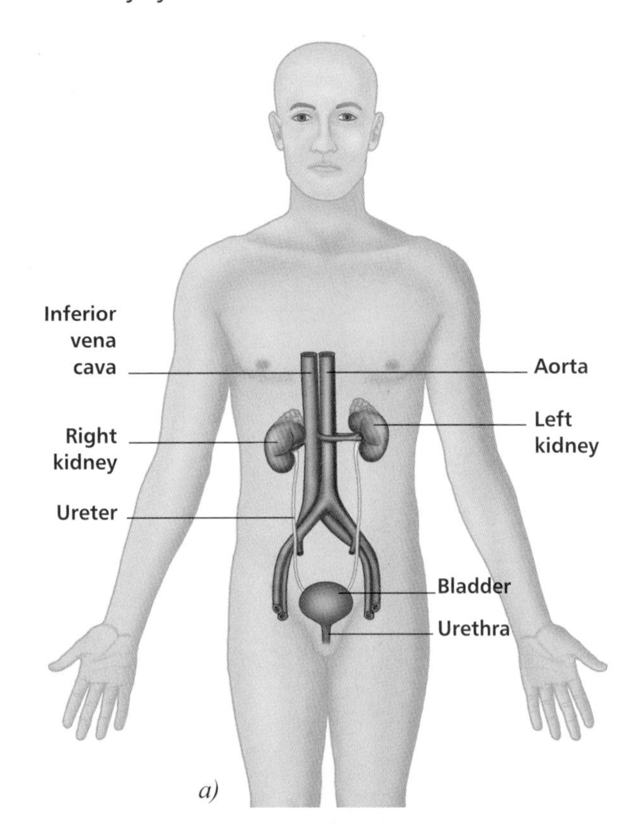

Figure 1.12: a) Excretory system, b) a skip, full of waste from a house, representing our excretory system, expelling waste resulting from our bodily functions.

Reproductive System

Living in our imaginary house is a family. The couple are architects and, working from their study, they design and build new houses. They aim to design perfectly built individual properties, of all shapes and sizes. Our own body has the capability to design and build new bodies, and the system that makes this happen is the *reproductive system*. Housed within the pelvis, the male and female reproductive systems are designed to make unique, individualised copies of ourselves, each having a recognisably human body, yet having distinguishing features and each different from every other.

You can see that by finding systems that exist in our daily lives, and relating them to anatomy and physiology, we can make sense of the structure and function.

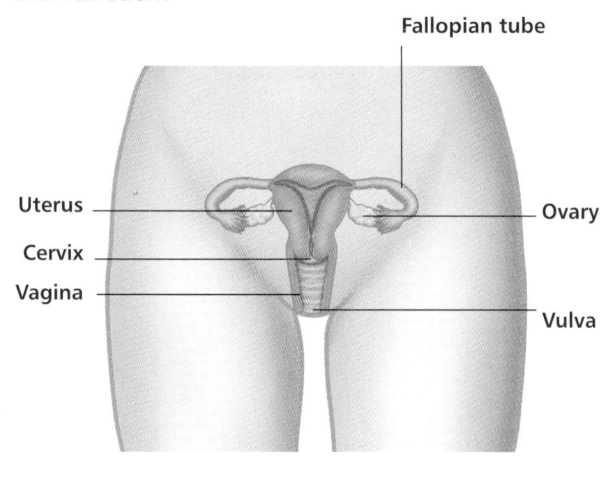

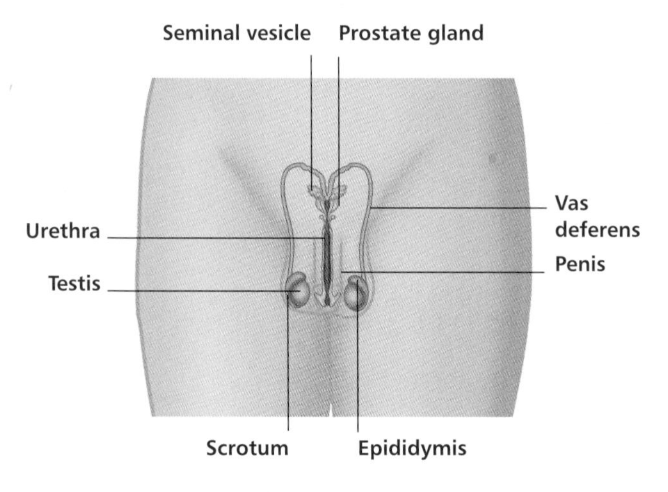

Figure 1.13: Reproductive system

The Cell

Looking from a cellular perspective, cells are the smallest unit of life and are the building blocks making up our body. Cells are formed in accordance with the genetic information encoded in their *DNA* and by signals from their environment. They are designed according to their functions and the structure of the organ in which they reside. Each cell contains different kinds of internal structures called *organelles*.

Let us take a trip around the cell, likening it to a factory, and thinking about the functions of each organelle.

The 'factory cell' is a structure that consists of sturdy cavity walls with doors for staff, vehicles and its manufactured products to go in and out. The shape of the building will be dependent on the size of the factory and its reason for being.

Cell Membrane

The outer walls are known as the *cell membrane*. Cell membranes are made from a double layer of lipid and protein molecules.

Cytoplasm

The factory floor of our factory cell is known as the *cytoplasm*, and makes up the entire contents of the cell except for the nucleus and the cell membrane. It contains lots of other structures that have key roles within the factory, with interesting names such as ribosomes, endoplasmic reticulum and Golgi bodies. More about them later!

Nucleus and Chromosomes

The *nucleus* is the 'head office' for the factory, where the people in charge tell the rest of the factory what to do. The managing director of the cell is the DNA: strands of specific molecules called *chromosomes* which take the responsibility of telling the rest of the organelles exactly what to do and when. The nucleus really is very bossy and can be thought of as the controller of the factory cell. It is usually a round structure but can take unusual shapes.

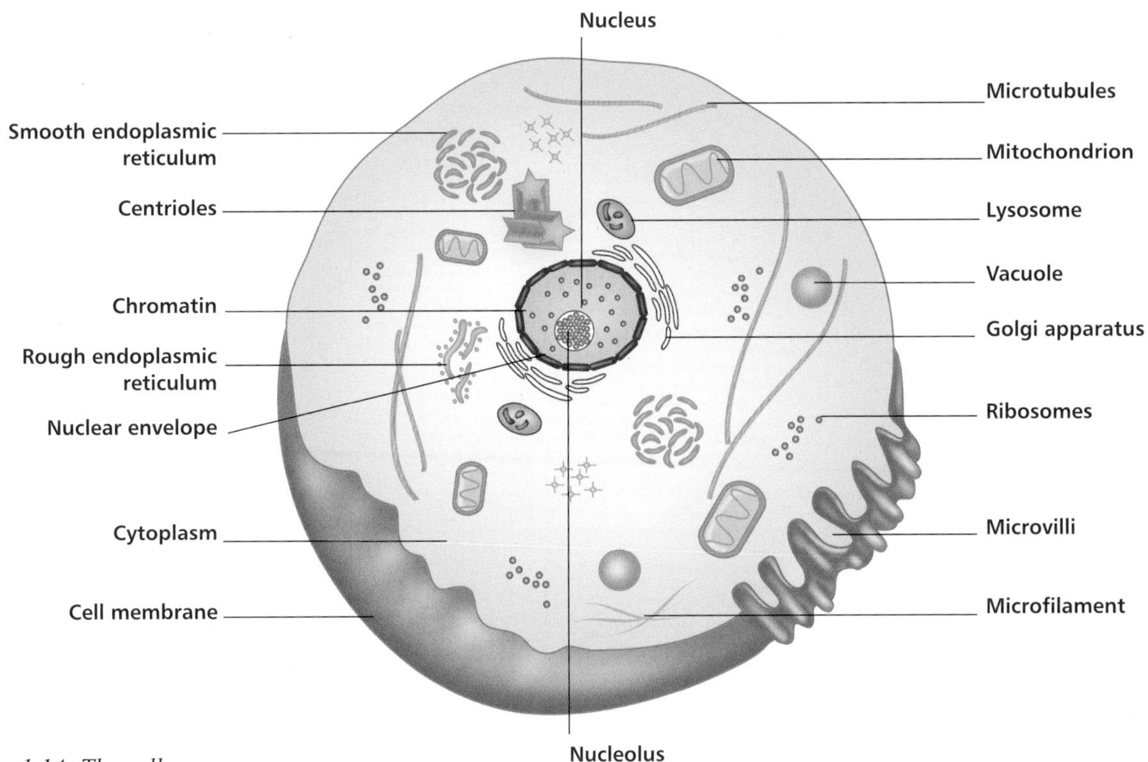

Figure 1.14: The cell.

Nuclear Membrane, Nucleolus and Nuclear Pores

Every managing director of a factory has an office, often with double-glazed windows or glass walls so he or she can see what is going on at the factory floor, and a door so important information can go in and out of the office. In our factory cell, instead of double glazing we have a *nuclear membrane* which consists of a double layer of membrane, and instead of a door we have a *nucleolus*. Its function is to produce *RNA* (ribonucleic acid) molecules within the nucleus and turn them into *ribosomes*, which are organelles that make proteins, using the plans found in another type of RNA made in the nucleus, the messenger RNAs transcribed from genes. When looking at the nucleus under the microscope, or at a photograph of a nucleus, the nucleolus is the paler area within it. The RNA strands are transported out of the nucleus via *nuclear pores*.

Rough Endoplasmic Reticulum and Ribosomes

The purpose of the factory is to manufacture products of high quality that are fit for purpose. Our factory cell has two main production areas. The area that makes proteins is the *rough endoplasmic reticulum*. It acts a bit like a conveyor belt and has *ribosomes* studded in it. Ribosomes are cheerful little factory workers shaped like snowmen, with a small head and a larger, rounder body. Think of little round people along a zigzagging conveyor belt and you get the picture!

Smooth Endoplasmic Reticulum

The other production area of our factory cell produces the fats and lipids that are needed, and stores calcium. It is more automated and does not have little people manning the conveyor belt so is much smoother in construction. This smoothly operating conveyor belt is known as the *smooth endoplasmic reticulum*.

Golgi Apparatus and Vesicles

Once a product has been made, it has to go to the packaging department, where it is packaged up ready for sending out to customers. Our factory cell has a packaging department too, called the *Golgi apparatus*, that modifies, stores and routes the products of the endoplasmic reticulum. The Golgi apparatus looks like a stack of flat sacks, all ready to be filled with the products of the factory cell. To transport the products within and out of the cell, they are packaged into special sacks (or in anatomical spelling, sacs) called *vesicles*.

Lysosomes

Every factory has a clean-up team of caretakers and cleaners to deal with spills and contamination, ready to pour on disinfectant to kill any germs in the spills and to empty the rubbish bins and to pick up litter and recycle it. Our factory cell has this in the form of *lysosomes*, which are like little bags of clean-up chemicals that destroy harmful bacteria and pick up any damaged organelles within the cell to recycle them.

Mitochondria

Factories need a powerhouse where energy is produced by burning fuel. Our factory cell has a powerhouse in the form of *mitochondria*. These actually look like mini-cells and have two layers of membrane: a smooth outer layer, and a folded layer inside. These inner folds are called *cristae* and mean that there is a bigger area for fuel to be burned. Fats and sugars are broken down and are used by the mitochondria to 'charge up' chemicals, in a similar way to charging up batteries with chemical energy. In our factory cell, the chemical energy is in the form of a molecule called ATP, which contains some stored energy that can be released when a section of it is removed.

Cytoskeleton, Microfilaments and Microtubules

The whole factory building is supported by a network of scaffolding and joists which holds up the building, and lifts, stairways and shafts that allow people to move around the building. Our factory cell has a *cytoskeleton* that is a network of tiny fibres and tubes called *microfilaments* and *microtubules*, which act as support for the cell and serve a variety of mechanical and transport functions.

Centrioles

Good manufacturing companies employ staff to keep an eye on the level of need for the factory's products in the market, and to alert the boss when it is time to expand. These people will oversee the growth of the company, and any future expansion and infrastructure needed for the opening of new factories. In our factory cell, these staff are called *centrioles*. Their role is to provide the structure of microtubules needed for the nucleus to multiply into two and separate into two distinct areas, ready for the cell to divide into two identical copies of itself. The process of division of a cell into two identical copies of itself is called *mitosis*.

Contextualise the Knowledge

Building on the foundations of anatomy and physiology knowledge requires the information to be made useful and memorable. Just looking at individual building bricks and cables and roofing tiles does not make you appreciate the intricate architecture of the finished house. Likewise, anatomy and physiology need to be brought into context in order to fully appreciate their intricacy and their use. For example, an unfinished house may not be attractive to look at, but each component needs to be carefully considered so that the finished product can function optimally.

When things go wrong in the body, the study of anatomy and physiology attempts to make sense of the faults or medical conditions. This study of disease processes is called *pathology*, and certainly makes physiology more interesting. After all, we remember when things go wrong in a house build, so why not use the pathology of the body to make physiology more memorable?

Every anatomy and physiology course will have a list of pathologies and medical conditions that are either covered by the course, or are part of the intended learning. Look these up at the beginning of your studies so you can begin to listen out for these medical conditions. If you actually know someone with the condition described, it makes it more memorable and certainly more real. By visualising the person concerned, you will understand the human aspect of the disease, rather than thinking of it as a textbook anomaly.

Foundation Stones – Finding a Level

How deep should our foundations in anatomy and physiology be? The depth of the foundations and introduction to anatomy and physiology really is dependent on how much prior knowledge the students have, and the extent that they will be using their anatomy knowledge. For instance, if a student has already passed a biology course, anatomy and physiology will be a gradual step up rather than a huge leap. For the student who has never studied any biology before, the concepts of cells and body systems will be a massive leap.

The depth of foundation is also dependent on the height of structure placed upon it. This is demonstrable by the need for more thorough groundings when extending anatomy for those learning at higher levels.

So, How Do We Know How Much We Need To Know?

Test Prior Knowledge to Work Out Entry Level

Working at the level required for the course undertaken, do a quiz or short test to find out how much you know about anatomy and physiology. Does this quiz make you interested and want to know more?

Regular Assessment to Check Learning

Regular quizzes, tests and assessments throughout the course keep the student keen and committed. These help work out if the learning has been adequate and meets the intended learning outcomes. They can be formally performed and the results recorded, or given informally as self-assessment.

Self-assessment at the End of Each Level/ Section

Testing oneself at the end of each module or section empowers the student to really acknowledge that they have learned exactly what they set out to learn. Many anatomical textbooks have online resources that contain quizzes and tests designed for the student to self-assess. Do use them!

Teachers must define intended learning outcomes carefully so that students can identify the level of information and detail required to pass the examination (see Chapter 7 – Test Anatomy).

Extending the Knowledge

Having constructed good firm foundations in anatomy and physiology, learning must be built, brick by brick, to give a strong and stable and permanent memory. How do we do this? How can we build on the knowledge?

Making connections between the structure and the functions of the organs, tissues and cells will make the information more real. By grasping basic concepts, and building on these, it can be seen that understanding concepts is more important than memorising lists of facts and names that have no meaning to the student. The concepts that need to be addressed early on in the learning of anatomy and physiology will lay a bedrock of knowledge that can be built upon with specialisms or with specific modules or chapters that look in detail at the application of these concepts. Conceptual learning encourages creativity and intrigue and hopefully passion!

Our subsequent chapters will examine how we learn, and how you as a learner can maximise your learning in the most appropriate way for you. We will give hints and tips about the best way of revising and preparing for examinations, how to own your anatomical knowledge by using reflection, and how to learn in some fun ways. Just as a house is constructed from ideas and plans and turned into a reality, so learning anatomy and physiology needs structure, planning and tools to create it successfully.

Some of the key concepts that are vital to grasp in order to form the bedrock and foundations of learning anatomy and physiology can be thought of as threshold concepts; once grasped and understood, these concepts can blossom and grow, building on each other to create a bigger and clearer structure that makes sense.

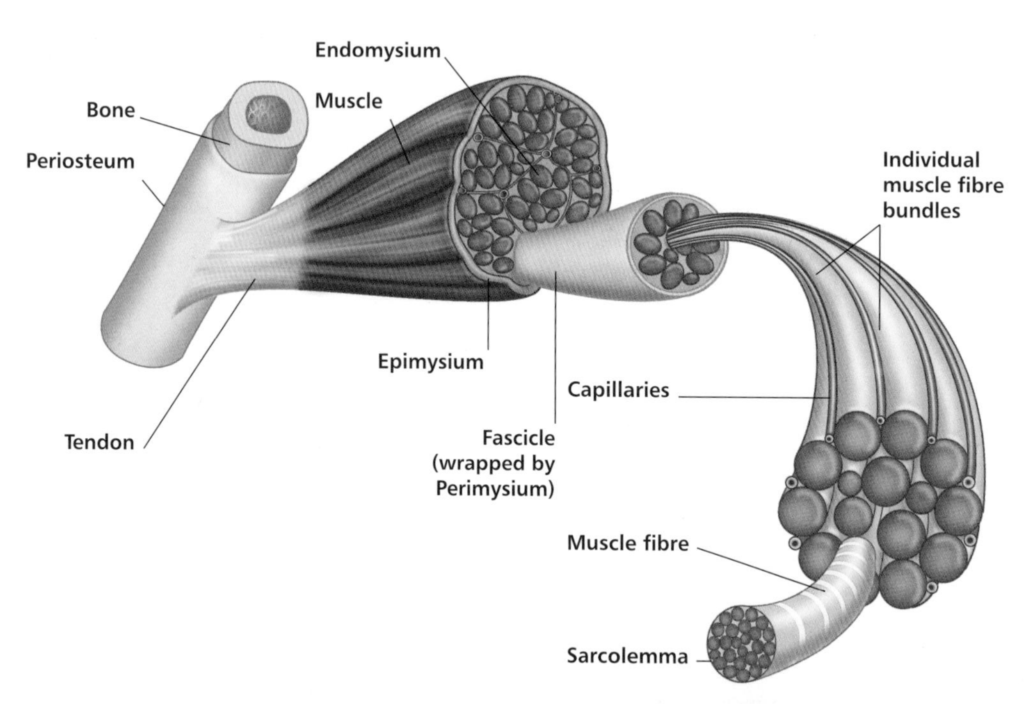

Figure 1.15: Bundles of muscle fibres are arranged in a similar manner to the arrangement of segments in an orange.

Key Concepts to Learn

How Cells Are Organised into Tissues and How Tissues Are Organised into Organs

- Voluntary muscles are made of bundles of fibres which contract when stimulated by the nervous system. They are arranged in a similar manner to the arrangement of segments of an orange: an outer skin, with segments surrounded by pith, and each segment consisting of tiny little pieces of orange flesh that lie in patterns, forming the integrity of the orange.

- In voluntary muscles, the fibres in the bundles are actually made up of lots of smaller bundles of fibres that are wrapped in connective tissue called epimysium. And the fibres in these smaller bundles are made up of even smaller bundles of fibres called fascicles wrapped in connective tissue called perimysium. And then each muscle fibre is wrapped in yet more connective tissue called endomysium.

- This 'bundle within a bundle' pattern even continues into the muscle cells themselves, where there are bundles of myofibrils, and those myofibrils are themselves made up from bundles of protein fibres called sarcomeres.

- The wrapping of each bundle in connective tissue means that it resists passive stretching, and also it distributes any force applied to it across the whole muscle. It also keeps the muscle fibres aligned in the best possible way in order for them to contract and move a limb or body.

Support and Connection

- Our bodies are not made as they appear in anatomy books, with parts that seem to clip together like a child's model. Instead we are a semi-solid 'soup' of collagen and connective tissue which is firm in some areas and runnier in others. Every part of our body is formed from this soup that has no beginning and no end, just a continuum. We need to be able to look at anatomy texts and pictures, and hold onto that thought of the collagen continuum, rather than compartmentalise and reduce our body to a slot-together kit. Our understanding of physiology and pathology and disease processes will make much more sense if we keep remembering this.

- Imagine a jelly mould shaped like a human body. Fill that mould with body parts made from jelly and pour in some more jelly and allow to set, making a model of a human completely made of jelly; all the organs float in a sea of jelly. Our bodies contain a type of tissue called connective tissue, which acts as the jelly in this analogy.

- *Connective tissue* does what it says on the tin. It is a type of tissue which connects virtually every part of our body together and fills the spaces between organs and tissues, providing structural and metabolic support for other tissues and organs.

- Connective tissue acts like mortar and cement in a house, forming a continuum which is connected. Just as mortar is made of a mixture of water, sand and cement, connective tissue is made from cells and a jelly-like gloop called *extracellular matrix* or *ground substance*.

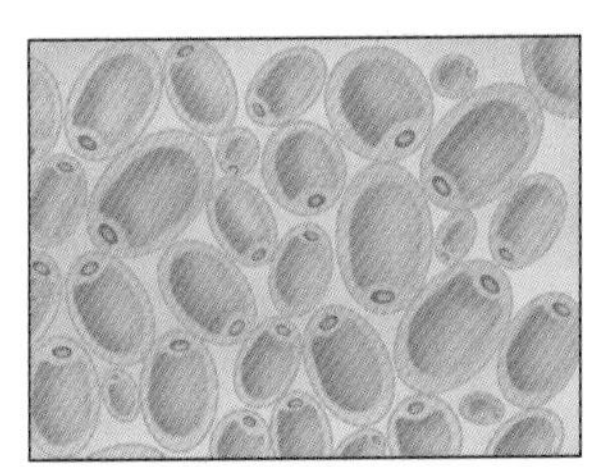
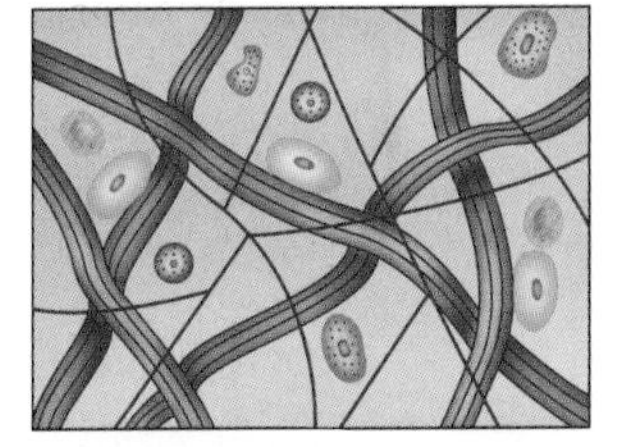
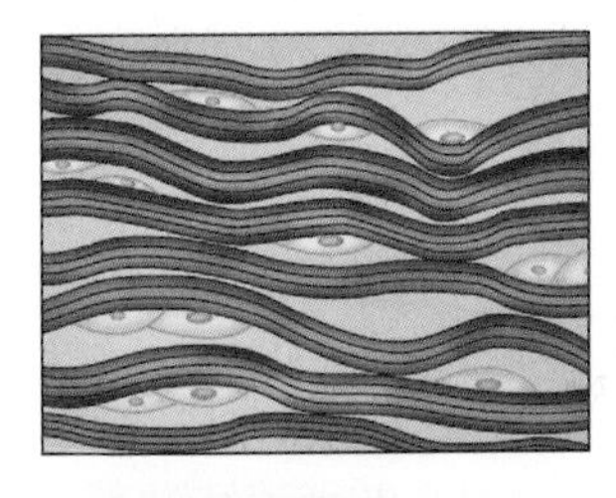
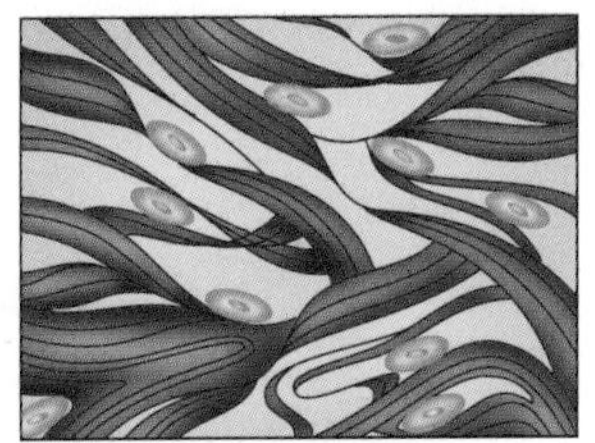
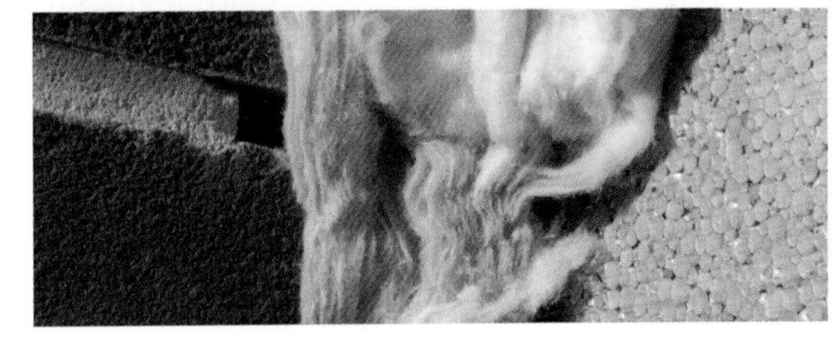

Figure 1.16: The various types of connective tissue in our body represent the mortar and cement of a house, holding the body together.

- Ground substance is clear, colourless and just like cement; it fills the spaces between cells, tissues and fibres. Its jelly-like properties are due to the shape of the molecules within it, which are large and link together, absorbing water like little sponges. The cross-linking of the molecules suspended in water gives ground substance amazing properties in that it resists compressive forces. When ground substance is mixed with structural fibres and cells, forming connective tissue, the ability to resist compression and to maintain shape means that bones remain strong, muscles do not tear, ligaments do not rip and we stand upright, yet remain able to move in a multitude of directions.

- A key concept of anatomy that needs to be completely understood and continually remembered is that connective tissue really does continue throughout the entire body. This means that there is a physical connection throughout our bodies, like an endless web of tissue. Remember the song, 'The foot bone connected to the knee bone'? Well it really is! In fact, the foot bone is truly connected to the stomach and to the eyes, even though they are separate organs. Considering this concept will make sense of injury patterns and disease processes.

- There are different types of connective tissue, depending on where it is and what its function would be. Some types are more solid, and function more for support, other types are shiny and smooth and would be for articulation (joint surfaces), and other types are loose and floppy and act as biological 'cling film' and 'bubble wrap', wrapping and packing our bodies into the correct space. The key concept to grasp is that all of these types of tissue are in fact made from the same ground substance and cells, but it is arranged according to need, with additional minerals and cells, depending on its intended functions.

Locomotion and Proprioception

- Movement is not just about bones and muscles. It is a coordinated combination of bones, muscles, nerves, brain, hormones, skin and other sense organs.

- Our bodies are designed to move. Movement is one of the properties of a living thing. Take some time to consider the complexity of processes which are needed in order to reach out and grasp an apple from the fruit bowl.

- Firstly, we are stimulated to look for food by a basic sensation of hunger, which is driven by a series of chemical, hormonal and neurological processes in the gut, blood and brain. Our eyes flick around the room, searching for a snack to quell the hunger pangs within us. Our spinal movement follows our eyes, and we slowly turn our head to face the fruit bowl. Tiny muscles which span two, three and four vertebrae contract and relax in a coordinated fashion which makes our head turn on our neck.

- We contract our anterior deltoid, coracobrachialis and brachioradialis muscles, lifting our arm, then these muscles are balanced with the action of the triceps muscle as we extend our arm. Our wrist and finger flexor muscles work to take hold of the apple and, with a combination of facilitation and inhibition of the arm and trunk muscles, we can retract our hand and arm and bring the apple to our mouth.

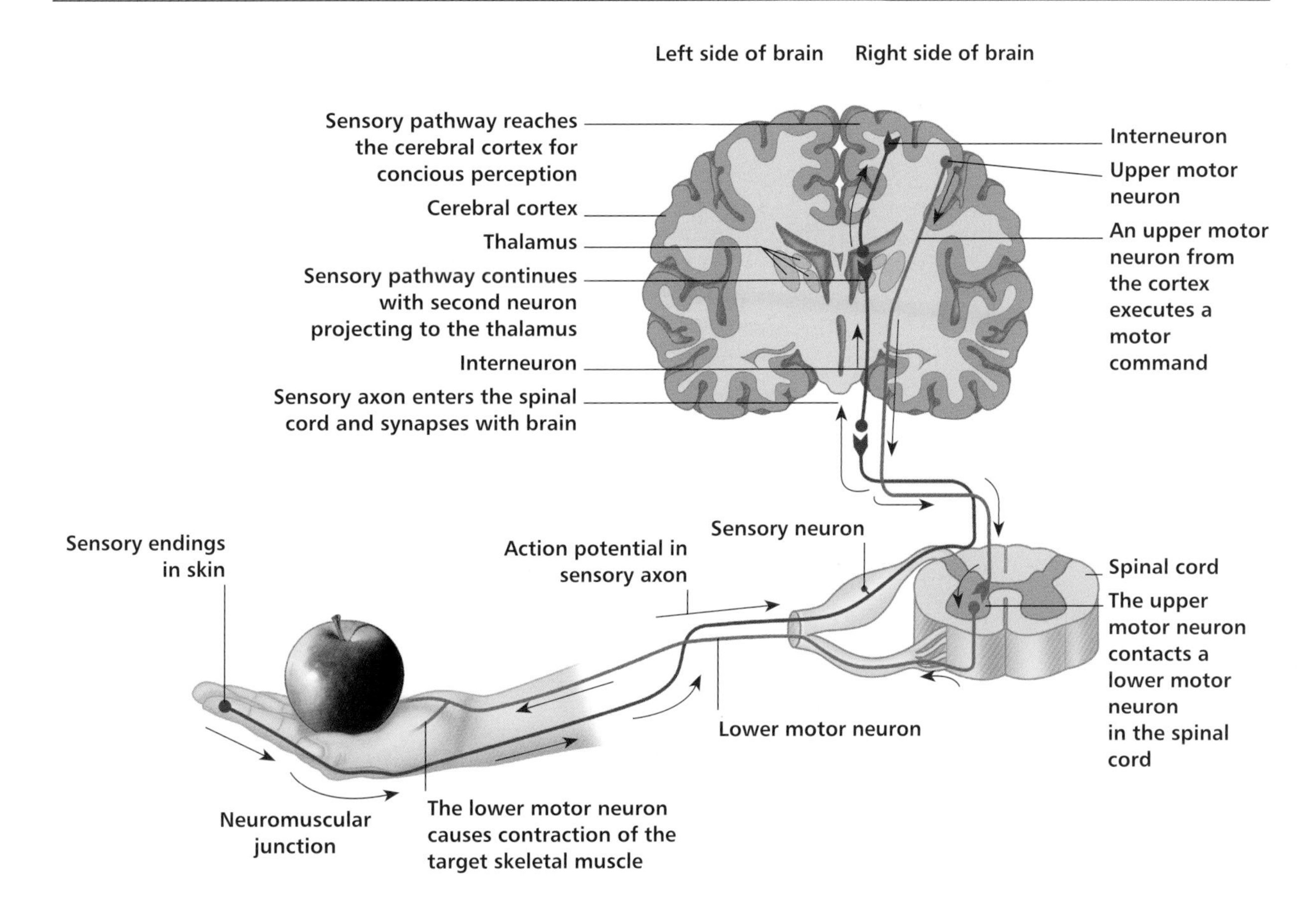

Figure 1.17: Muscle action is controlled by the voluntary nervous system.

- All of the muscle action is controlled by the voluntary nervous system: the peripheral nerves and the brain. However, our position in space and our sense of location are controlled by an automatic part of our nervous system, proprioception. Without this sense we would not be able to move in a controlled manner at all, and be either motionless or uncontrolled. Proprioceptors are nerves which have little sensors called sensory nerve endings in the joints, ligaments, muscles and skin that detect stretch and pressure. Our brain is truly amazing as it can process all this information, and piece it together to give a complete picture of where all our body parts are. It is proprioception that allows us to know that we have a hand reaching out for an apple. Try it. Reach out for something. Now close your eyes. You still know that your arm and hand are outstretched. Without proprioception, reaching out for the apple would mean that you topple over as the counterbalancing muscles that contract to prevent uneven weight distribution would not work. Likewise, the feedback from our fingers as we pick up the apple would not be felt and interpreted, so we might either drop the apple, or grip so tightly that the apple is bruised and dented.

- So movement is not just a set of muscles, bones and joints. It is a *whole body* system of integration and if any part of it is dysfunctional, then movement restrictions will occur, and disease and disuse may result.

Exchange and Diffusion

- Body fluid is not just called body fluid – it depends where it is in the body!

- Our body's fluid has been absorbed from our diet by the *digestive system*. It flows around the body in *blood vessels* under the name of *plasma*. It leaves the heart and passes via the aorta to the pulmonary artery to the *lungs*. It then flows back to the heart via the pulmonary vein. After being pumped back out of the heart to the *aorta*, fluid passes around the body through *arterioles*, and into *capillaries*. The heart is controlled by a combination of nerve activity controlled by the *nervous system*, and the action of hormones controlled by the *endocrine system*.

- Some of the plasma in the blood passes through the *kidney*, where excess fluid is pushed through the *Bowman's capsule* and becomes *glomerular filtrate*, and if required is fed back into the blood circulation at the *loop of Henle*. Unwanted glomerular filtrate is passed out of the body as *urine*. This is known as the *renal system* and is controlled by *hormones* of the *endocrine system*.

- As the blood enters the capillaries of respiratory system tiny amounts of it move out into the air-side of the capillaries. Some of this remains to moisten the lining of the lung enabling the exchange of gases to occur, and some of the water is lost as water vapour when we exhale. This is why we steam up a cold window when we breathe on it.

- The capillary network of our blood system supplies our *skin*. Some of our excess fluid will move from the blood to our sweat glands, where it will be excreted as *sweat* when the nerve endings in our skin recognise that we are too hot, in an effort to cool us down.

- Some of the plasma squeezes and diffuses out of the capillaries and into the space around the tissues and cells and is now called *interstitial fluid*. This bathes the cells with important nutrients that they need to function. Some of the unwanted interstitial fluid is able to diffuse back into the capillaries, where it is known as plasma again, and is taken back into the venous circulation, returning to the *vena cava* and into the heart. The rest of the interstitial fluid is unable to go into the capillaries, so drains and diffuses into the second circulatory system, the *lymphatic system*. The fluid is now called *lymph* and passes through *lymphatic vessels* containing bug-fighting cells and *lymph nodes* of the *immune system*, which clears and cleans the lymph, before it returns to the blood circulation at the thoracic and subclavian

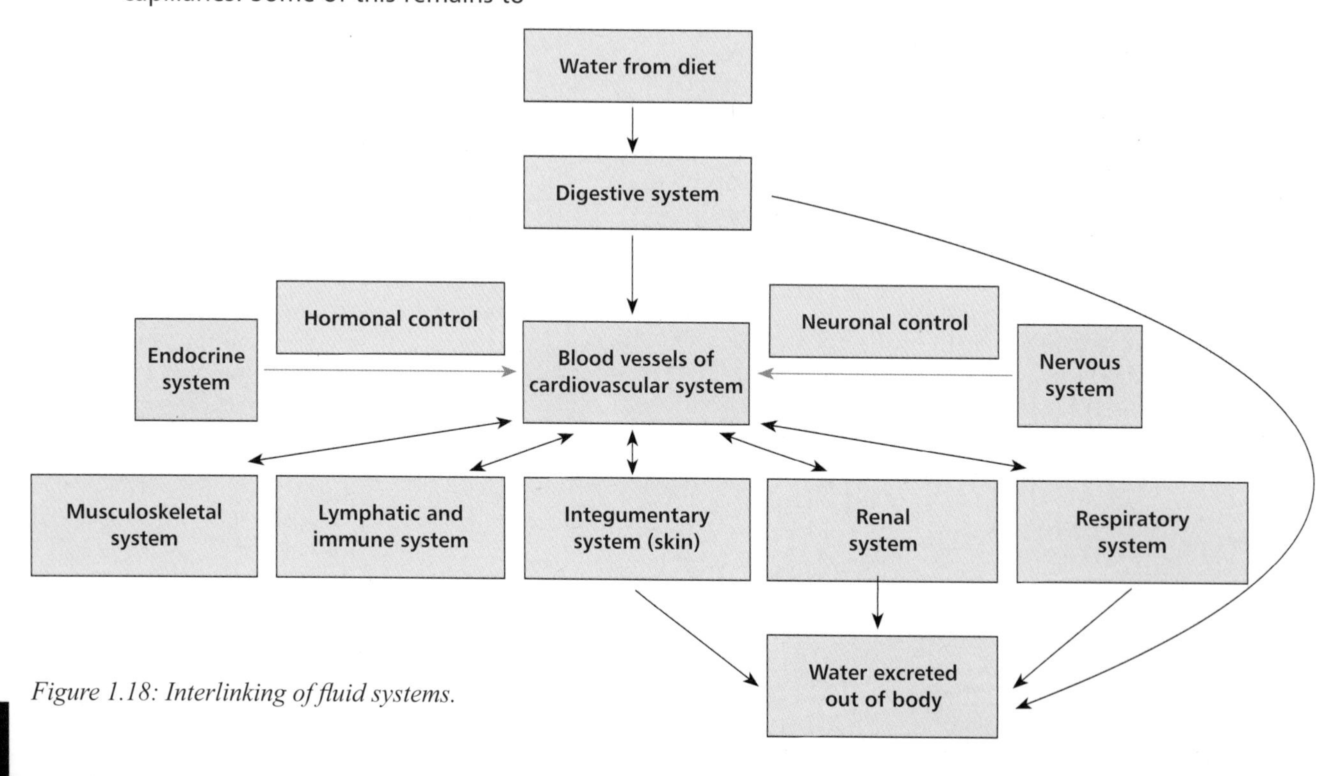

Figure 1.18: Interlinking of fluid systems.

ducts. There isn't an actual pump to make lymph move around the body – instead it relies on muscle activity, the suction action of the diaphragm and the stretch of the skin by movement of the body, which pulls lymph from the tissues.

- So, movement of fluid around the body involves:
  1. Digestive system
  2. Cardiovascular system
  3. Respiratory system
  4. Integumentary system (skin)
  5. Lymphatic system
  6. Musculoskeletal system
  7. Immune system
  8. Renal system
  9. Nervous system
  10. Endocrine system

- If any of these systems is not working correctly, then specific pathology occurs causing disease; the type of disease is dependent on which systems are affected.

Integration of Form with Function

- The shape and form of cells of the body are designed with their function in mind.

- One of the parts of the digestive system within the *small intestine* is called the *ileum*. One of its functions is to absorb nutrients and to send the nutrients via the circulation into the liver to be processed.

- The surface of the inside of the ileum looks like a microfibre mop! It has lots of tiny fingers that stick out, and these are called *villi*. The function of these is to increase the surface area of the inside of the ileum, increasing the amount of nutrients that the ileum can absorb. If it were flat, it would have to be incredibly long and our bellies would have to be absolutely massive!

Figure 1.19: The surface of the ileum looks like a microfibre mop.

- Now take a look at the shape of the cells that line the ileum, forming the *mucosa*. Most of the cells here are absorptive cells, which are closely packed rectangular cells and have an unusual cell membrane on the hollow side of the ileum (the *lumen*). Here, the cell membrane is shaped like tiny little villi, and is known as the *brush border* or *microvilli* because of its shape. The brush border replicates the shape of the inside of the ileum, right down to cellular level, giving a much bigger surface area for absorption than if the cells had a flat edge.

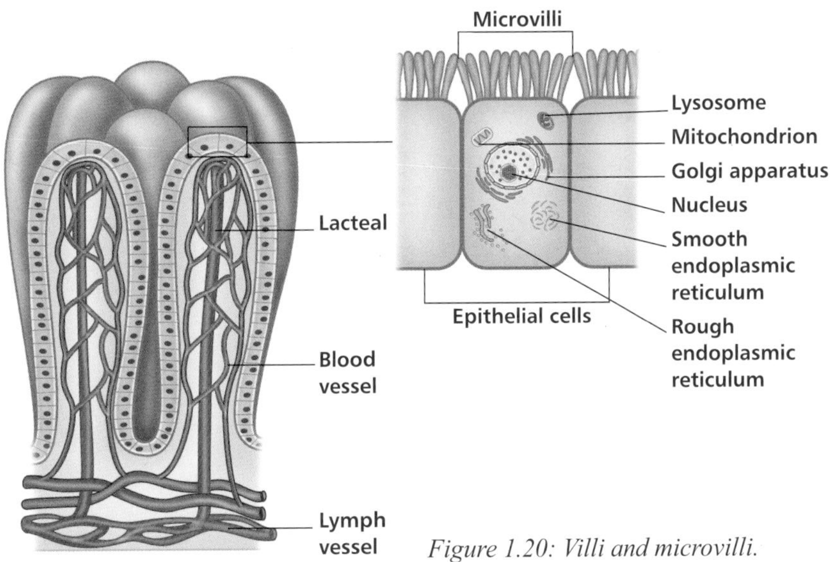

Figure 1.20: Villi and microvilli.

Taking the Information From Conception to Grave

The body systems are both interconnected and interdependent, yet need to be taught individually in order to create the foundation for learning. Deepening the understanding of the systems can be achieved by looking at the connections between them and by looking at them from various perspectives: from conception and the embryology of the body systems, the effects of growth and ageing, and also the effects of disease and pathology upon the systems and their tissues. Considering anatomy and physiology from these aspects will give a broader overview of the subject which is more interesting, more relevant, more enduring and much more memorable.

Embryology of the Anatomical Structure/System

Understanding the processes of the early beginnings of the body's systems helps one to understand the interrelationships between them, and also helps to explain some of the diseases and pathologies that emerge in older age. The organisation of the cells into specific areas which then develop into specific tissues and organs at such an early developmental age is truly amazing, and needs to be studied carefully and with awe.

Causes and Effects of Disease on the Anatomical Structure/System

An interesting part of the anatomy and physiology course is the study of pathology. This is defined as the study of the causes and effects of disease. It can be studied at a cellular level or at a systems level, at a microscopic level or at a gross visual level, or even at a biochemical level. The most memorable way of learning and understand the effects of disease is to relate all of these approaches to real cases or to published case studies. The human element will be very visual and will be more experiential and therefore more easily remembered.

Pathology is a massive topic, and can best be learned when the basic foundation of anatomy and physiology has been grasped. An anatomical appreciation of 'normal' is essential before tackling the 'abnormal'.

Effects of Growth on the Anatomical Structure/System

A characteristic of living things is that they grow. As the human body grows from a fertilised ovum, incredible changes in both organisation and structure occur. Once the baby is born, it continues to grow in size until adulthood, its structure being dependent on its needs and abilities. An example of this is the skeletal system. As a baby, this is mainly cartilaginous – after all, imagine giving birth to a fully calcified baby – it would fracture and break when squeezed out through a vaginal delivery! The baby's cartilaginous skeleton is able to take the pressures and rigours of labour and birth by bending and flexing yet returning to shape afterwards. As the baby ages, and its muscular and neurological control becomes more coordinated, bones begin to calcify and become stronger to take the load of walking and locomotion. The spine develops curves to give shock absorbency, strength, flexibility and balance as the baby becomes upright, walking on two legs.

We grow not only in height. Our body is constantly growing to restore tissue that needs replacing owing to damage or disease. Our skin cells are constantly being worn away by our environment, so need to be replaced. Our blood cells are constantly produced in the bone marrow to keep up with the turnover of cells destroyed by the spleen. We are simply NOT the body we were three years ago!

Effects of Ageing on the Anatomical Structure/System

The ages we go though in life dictate our anatomical structure according to its changing function and are closely linked to the effects of growth. When considering the effects of ageing on anatomy and physiology, do not plunge straight into old age – consider the other stages of ageing too:

- Babyhood
- Childhood
- Adolescence and puberty
- Adulthood
- Middle age and menopause
- Old age.

And the systems and functions that are affected by changes with age include:

- Hormone production
- Immunity
- The skin
- Sleep
- Bones, muscles and joints
- The breasts
- The face
- The female reproductive system
- The heart and blood vessels
- The kidneys
- The lungs
- The male reproductive system
- The nervous system.

The types of changes due to ageing that occur at a cellular level are:

- Atrophy: cells shrink and reduce in number.
- Hypertrophy: cells enlarge in size.
- Hyperplasia: cells grow in number.
- Dysplasia: size, shape or organisation of mature cells becomes abnormal.
- Neoplasia: the formation of tumours, either cancerous (malignant) or noncancerous (benign). They often reproduce quickly. They may have unusual shapes and abnormal function.

The reasons that tissues and cells change with age have yet to be fully understood; some changes are due to defects in the genetic code within the DNA in the nucleus, others are due to exposure to chemicals or ultraviolet light, and others are due to wear and tear of the body. Viewing anatomy and physiology from the standpoint of the process of ageing imparts a different understanding of our bodies and their structures.

Final Thoughts ...

We recognise that the success of learning anatomy and physiology hinges on the learning of concepts that can be applied, rather than memorising facts. The process of learning to use information is as important as the concepts themselves. Using understanding to explain and make connections is a more useful long-term lesson than is memorisation. Anatomy should be presented and learned as a dynamic basis for problem solving and for application in the practice and delivery of quality healthcare (Miller, 2002).

References

Miller, S.A., Perrotti, W., Silverthorn, D.U., Dalley, A.F. and Rarey, K.E. (2002). From college to clinic: reasoning over memorization is key for understanding anatomy. *Anatomical Record* **269(2)**: 69–80.

2 Speak Anatomy

The study of human anatomy has received much attention. The fascination with understanding the workings of the human body dates back many centuries. Students of anatomy have always been intrigued, confused and often bewildered with the complexity of the human body and the way in which component parts work together to produce an individual capable of movement, cognition, behaviour and inquiry. For many, the human body is a work of art, a perfect machine that can adapt to and control many environments. For the student, however, the learning of anatomical concepts, applications and terminology can be troublesome and problematic. How best can anatomy be learned and taught? This chapter deals with understanding the language of anatomy and provides clarity on how best to develop this language for both learners and teachers. Through this chapter we connect language with learning to enable the reader to make sense of the words and text of anatomy and physiology.

I often encounter students sitting in hallways, on stairs or in libraries with heavy anatomy texts, writing furiously in multicoloured pens, underlining key terms, sticking Post-it notes all over the chapter, and drawing and labelling diagrams. What exactly are these students learning? Are they intending for knowledge to passively flow from the page into the cortex of their brain and be stored there for future use, or is there an active method that is being implemented to ensure that the learning content and knowledge has meaning and can be understood and applied? Anatomy, like any other clinical study, is punctuated with complex terminology, concepts, applications and structural analyses. For some the science is exciting and can be readily understood and applied, but for others this learning can create a threshold, a troubled experience and mysterious learning space, one which needs to be navigated.

I often ask anatomy students how they arrived at their answers, and far too often I receive the same answer – that is what it says in the book. This clearly illustrates that learners tend to bypass genuine understanding and are often comfortable having an answer, even if it lies beyond their defence. We tend to be moving into a world where we rely on finding answers at the expense of asking the right questions. Anatomy is not always about finding answers. At times, it may be more useful to ask different types of questions to influence new and innovative ways of thinking and knowing. We acknowledge that different cultural biases exist in language development and origins of words. This chapter is about how the understanding of terms and concepts influences the development of anatomical knowledge.

Let's Start at the Very Beginning ...

So there you are, first minute, first session and possibly first exposure to anatomy. The text you use may be fairly detailed, depending on the reason for you having to study anatomy. Remember, many students have taken this journey before. There have been many first days and first experiences with anatomy. Identifying and relating anatomy to everyday living brings meaning to the content and explains how the body functions to sustain life.

Anatomy, as a study, is often defined by specific objectives and tailored to meet set learning and teaching outcomes. These could be developed by institutions of learning, or set by awarding bodies who require core knowledge understanding and application. Whatever the reason, or administration, of the programme in which you learn anatomy and its allied sciences, the bottom line is that the anatomical content needs to be learned.

You may begin by browsing through the chapters, examining a few words, looking at diagrams and pictures, or reading text blocks. Your preferred learning style will dictate the method you adopt.

Along your initial journey, you may encounter an internal fear, a realisation that the content is greater than your current understanding, or ability. You may even think you have enrolled in a foreign language class. Some of the words may appear to be Greek or Latin – you need not worry, they probably are. You may ask yourself why are you here or how will you ever digest the volume of work you need to learn in order to satisfy the examiner.

Fear of encountering the unknown creates an unparalleled drive to develop new understanding and application. When we study we often relate new content to what we already know, or have experienced. We constantly compare and contrast knowledge and learn to accommodate and integrate new information. Our ability to learn and adapt makes us unique. How often have you had to read and then reread content to develop an appreciation for understanding? As we gain more experience and build our knowledge we learn to read with a renewed focus and perspective. It is almost like taking a photograph of nature from a different angle or perspective. With time, effort and experience, we learn to see things we failed to see the first time we read the work. To learn about the body we need to start at the beginning.

Lost in Translation – The Language of Anatomy

My students found this old arm bone – I think it is humerus

With this in mind, let's begin by considering anatomy as a language. A young child tends to pick up languages quickly and effortlessly and is able to strike up conversation, even if the grammar needs some fine tuning. So, if young children can learn languages quickly, so, too, can you. In fact your brain is already wired to learn a new language. Learning a language is not about constant repetition, it is about listening, anticipation, practice and identifying the key words. Instead of trying to memorise concepts and applications, by learning the main ideas of language, you will soon learn to recognise words or parts thereof, and begin to organise your thinking in a way that facilitates your learning. Anatomy is often described as a visual study, and by visualising structure coupled with a formation of ideas and anticipated responses, you will be able to master concepts without spending long hours reading texts.

When learning and teaching anatomy, it is important to learn or teach the right word in the right way. The greatest thing about learning anatomy is that one does not need to know all the terms to start speaking anatomy. Learning anatomy from scratch is virtually impossible as you already know a vast amount of English or other language words. This is because languages are built on cognates, i.e. words we already recognise from our native language. Many mature language learners use the adage, 'I'm too old to become fluent.' But adults can be better language learners than children. Research has found that, under the right circumstances, adults show an intuition for learning languages.

If you were asked to build a house and had no prior knowledge of house construction, where might you begin? You could start by reading books or texts on building, watching videos, or speaking to people in the know. The study of anatomy follows the same thinking process. Speaking to someone who knows about building may be a more effective start. If we translate this idea into learning anatomy, it is often useful to speak with learners and learn with and from them. Anatomical concepts that are simply transcribed in text may often be difficult to comprehend. If you are able to use your knowledge of language dissection you may begin to appreciate the clues in the word. It is like having a magic key to unlock the mystery and often misery that may surround the word, idea, application or concept.

Just like building a house requires a solid foundation, so, too, does studying anatomy. It is important to ensure that the content learned is strengthened through applications and use in different situations. This chapter is about unlocking your potential and learning to navigate the boundaries of the human body. It is not about sitting down and trying to digest volumes of work, but more about being smart about the way in which anatomy comes together and comes to life. Like any study, the study of anatomy is more about your individual experience of learning, and how collectively we can influence the way we learn and apply our knowledge. Use this chapter to guide you through the recognition of anatomical terms, the mastery of the language of anatomy, and your unique ability to decipher meaning from anatomical concepts and terms.

The journey of learning anatomy as a language begins with an inspection of words, phrases and concepts. Like a detective trying to decipher meaning from clues, so too does the learner, in his or her quest for understanding and knowledge. To help prepare for the journey of learning, it is important to consider some key questions:

Do we speak the same the language?

Where do we, or should we, begin?

In any language there is the connection of letters to form words and words to form sentences. Before we learn words, we learn to recognise characters in the form of letters and symbols, and associate letters with certain words, e.g. 'a' is for apple. In the study of anatomy, the words we use often stem from anatomical positions and relational terms. Consider the images in Figure 2.1.

The anatomical position is commonly used to teach students about the relationship between structures, by focusing on the body in a forward-facing stance. With this stance engrained in the brain, students learn about the imaginary midline which dissects the body into perfect left and right halves. I used to have students bake gingerbread men and bring these into class as learning aids. We soon learn that this anatomical position is unique in describing the body and is almost the body's map for describing anatomical structure. This enables us to develop an anatomical communication system, a way of developing our understanding through a common pathway. We learn to punctuate our thoughts with correct terms and descriptions and standardise our language construction.

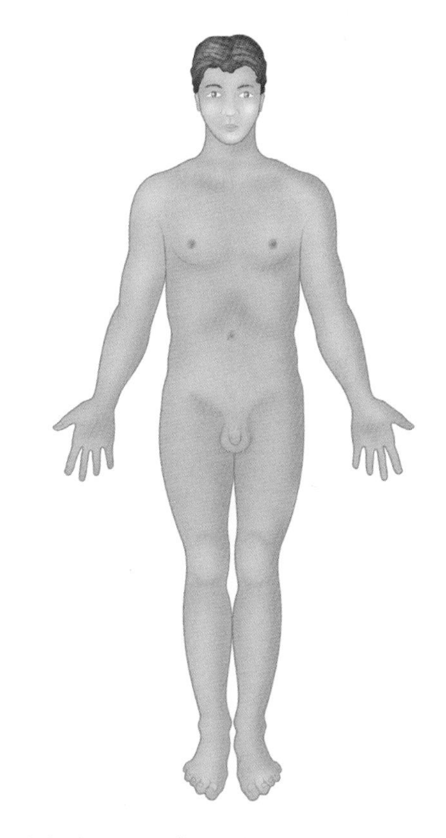

Figure 2.1: Anatomical position.

Developing Our Communication: from Novice to Neuron

We now focus on one term, the term used to describe structures that lie close to the midline. That term is *medial*. If we shuffle the letters in the word we discover that the letters **m e d i a l** could be rearranged, if we omit the 'a', to create the word **midle** – this creation gives us a clue to the description of the word, as it sounds like *middle*. If this is applied to structure, e.g. vastus medialis, we can separate words to position structures within the body. In the example of vastus medialis, the term medial appears. If we were unsure about the other word, vastus, we at least can deduce that this structure is likely to appear towards the midline of the body. This association of words and relational terminology is important in shaping anatomy as a language.

The table listing anatomical terms provides some examples of word and term associations. Use the table to consider the key learning dynamics and then extend your understanding by adding more terms to the table.

Taking the ideas presented in the table, we can now begin to create some interesting learning cards to help animate anatomical concepts. A few ideas appear below. Once we understand and use the language effectively, the study of structure in relation to function may become more familiar.

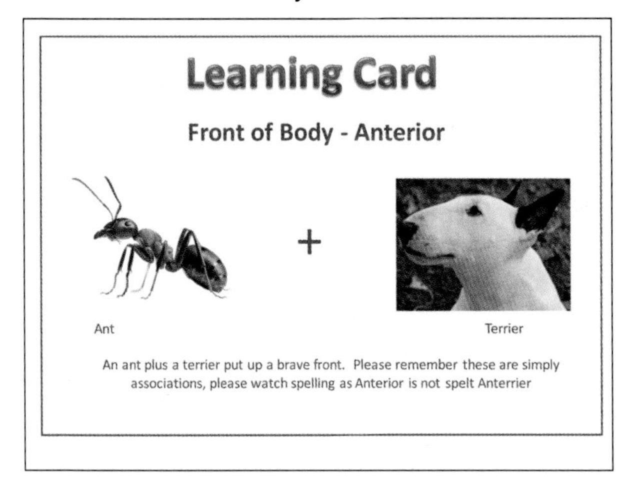

Figure 2.2: Learning cards.

Anatomical term	What it means	Learning point
Anterior/ventral	Front of body	Associate *ant* with *terrier* as a front line
Posterior/dorsal	Back of body	Think of the buttocks as a *post* box
Superior	Towards the head	People who think they are superior have big heads
Inferior	Towards the feet	People who think they are inferior look down on themselves
Proximal	Towards the origin/point of attachment of a structure, or the median line of the body	Consider that *pros* are on top of their game
Distal	Away from the origin/point of attachment of a structure, or the median line of the body	Distal is some *dist*ance away
Abduct	Move away from the midline	Someone who has been abducted has been taken away
Adduct	Towards the midline	Double *dd* in the word to suggest 'down down', or a*dd*ed to the body
Flexion	Reduce angle of a joint	To flex means to bend
Extend	Increase angle of a joint	When we extend we increase the area

What's In a Name?

Learning a language can be exciting and problematic at the same time. This section deals with word associations and identifying prefixes to words that inform concepts and terminology. Often, knowing where a term originates, or associating set letters with words, can assist in learning anatomy and physiology.

The table provides an overview and introduction to some of the more common prefixes and associations. Use the table to help you understand the journey through anatomy and physiology. It would be useful to update the table, by adding further prefixes and word associations, when you develop your learning and knowledge further.

Prefix/association	**Body reference – derivatives**
Antebrachi(o)/antebrachial	Forearm
Brachi(o)/brachial	Arm
Bucc(o)/buccal	Cheek
Cardi(o), card/cardiac/cor	Heart
Carp(o)/carpal	Wrist
Cervic(o)/cervical	Neck
Chondr(o)/chondral	Cartilage
Cost(o)/costal	Rib
Crus/crural	Leg
Cubito/cubital	Elbow
Cyt(o)	Cell – cytology/cytoplasm
Ec, ect(o), ex(o)	On the outside
End(o)	Within – endomysium
Faci(o)/facial	Face
Femor(o)/femoral	Thigh
Haem(o)	Blood
Hepat(o)/hepatic	Liver – hepatitis
Hist(o), histi(o)	Tissue –. histology
Hyper	Above (hypertension = high blood pressure)
Hyp(o)	Below (hypotension = low blood pressure)
Labio/labial, labral	Lip
Lumb(o)/lumbar	Lower back
Mandibul(o)/mandibular	Lower jaw
Mani/manual, manus	Hand –manicure
Maxill(o)/maxillary	Upper jaw
Ment(o)/mental	Chin
My(o)	Muscle
Nas(o)/nasal	Nose
Neur(o)/neural	Relating to nerves
Ocul(o), orbito/ocular, orbital	Eye
Oro/oral	Mouth
Os	Bone
Ot(o)/otic	Ear
Ped(o), pedi/pedal	Foot – pedicure/pedal
Peri	Around – periosteum
Pneum(o), pulmon(o)/pulmonary	Lungs
Ren(o)/renal	Kidneys
Sub	Under, e.g. subscapularis, a muscle under the scapula
Thorac(o)/thoracic	Trunk

One area that many students struggle to master and often find difficult to learn is that of naming skeletal muscle. Naming skeletal muscle is an excellent starting point for considering how names are shaped and formed. Often, the clue is in the name and knowing this can be a useful strategy for understanding the musculoskeletal system. Take, for example, the sternocleidomastoid muscle – this muscle is built on three anatomical structures which not only constitute the name but, further, signpost important bony landmarks and attachment points. *Sterno* relates to the sternum, *cleido* to the clavicle, and *mastoid* to the mastoid process of the temporal bone. Once students learn the bone names and their processes and points for muscular attachment, positioning the muscle on the skeleton becomes more manageable. The name of a muscle can indicate a range of useful information, such as location, position, number of heads, action, shape and size, to name but a few.

How muscles are named	**Examples**
Shape	Deltoid (triangular), rhomboid
Size	Maximus, medius, minimus
Action	Flexor, extensor, abductor
Number of heads	Bi (2 heads), tri (3 heads), quad (4 heads)
Attachments	Sternocleidomastoid – sternum, clavicle, mastoid process of temporal bone
Location	Femoral, radial, ulnar
Fibre direction	Rectus, transverse, oblique

Sequencing and Ordering Words – Problems with Learning a Language

The Rubik's cube is an interesting learning puzzle that envelops the concept of challenge. The cube has become the prize and bane of many. The cube presents a challenge, one which requires solution. Some use this challenge as a mastery technique, whilst others become perplexed, confused and frustrated by trying to solve the cube puzzle. This confusion and frustration can often lead to cheating, i.e. an attempt to peel off the colours and reposition them to solve the cube. Anatomy as a language may present the same frustration initially. Learning new words or learning that a phrase could have multiple meanings is equally frustrating. Unlike the cube it is difficult to peel off names and reposition them. The cube phenomenon teaches us patience and perseverance. Frustration could lead to a defeatist attitude. We need to remember that learning takes time and that the quality of learning is just as important as the quantity of learning, if not more so. The Rubik's cube teaches that, by fine-tuning, revisiting and viewing a problem from different angles, one is better equipped for mastering technique.

Figure 2.3: Rubik's cube.

Consider whether you are a superficial or deeper learner. Are you one who skims the surface, or pauses to admire and appreciate the content beneath? The poem below is a wonderful example of how we can get by without fully understanding what it is we are reading and/or learning. In reading 'Jabberwocky', we learn to enter a world of the foreign, mysterious and highly nonsensical language relating to the cognition of the poet.

> *'Twas brillig, and the slithy toves*
> *Did gyre and gimble in the wabe;*
> *All mimsy were the borogoves,*
> *And the mome raths outgrabe.*
>
> First verse of 'Jabberwocky' – Lewis Carroll, 1872, *Through the Looking-Glass*

Questions:

1. What did the slithy toves do?
2. Describe the borogoves.
3. Why were the mome raths outgrabe?
4. Explain the relationship between the slithy toves and mome raths.
5. What is the mood of the poem?

Questions 1 and 2 demand nothing more than finding words in the poem to generate an answer. Little understanding of the poem is required to answer these questions. In anatomy examinations/ assessments, questions are often developed to allow learners the opportunity to find the answers within the examination paper itself. To challenge learners and their learning, questions 3–5 require a deeper understanding and analysis of content.

The poem may do nothing more than invoke feelings of amazement and amusement. We may even go as far as being able to answer simple questions about the poem, despite not knowing what it is the poet is writing about or contextualising. Similarly, in the study of anatomy a superficial understanding may suffice in navigating basic understandings of concepts but is limited in enabling the learner to develop a deeper appreciation and understanding of the concept.

At different times language may confuse, inform or shape our understanding. What remains important are the ways we apply our understanding and develop our conceptual tools, and how we continue to develop our knowledge through reflection and consolidation (refer to Chapter 8, on the anatomical toolkit and reflection).

The study of anatomy focuses on the structural alignment of body parts. Using this theme we can further consider the learning of anatomy as the structural alignment of letters and words. For example, *myo* refers to muscle, *card* to heart, *ren* to kidney, and so on. Anatomy is an applied study and one that allows the learners and teacher to present information in a visually exciting, meaningful and innovative way. The success to learning anatomy lies in our ability to recognise the clues in the words; examine their meanings and origins; and begin to construct understanding of the concepts, principles, protocols and processes we intend learning. Like all languages, once we understand a few rules, can align letters to form words, and then deconstruct words to better appreciate meaning, we begin our process of mastery.

Anatomy includes more than just words and terms, it is a language within a language. In fact, anatomy is a unique communication network. Consider the way that blind individuals learn through touch. Anatomy encourages learners to palpate structures and differentiate different structures and bony features. Learning through feeling is a wonderful technique to enable learners to internalise structures and develop a much richer picture of what it is they are learning. Through the study of anatomy we don't only consider the structure of the body, but, more importantly at times, the shape too. Students need to be able to consider the shapes of bones, the connection between structures, the lines of pull of muscles and the arrangements of structures within the body. In language, we can often change meaning by moving a comma or punctuation mark. In anatomy, confusion of terms could lead to the creation of creative names for structures that do not appear in any text or applications.

To further elaborate, spelling of words is critical in anatomy. For example, if we spell *ileum* with an 'e' we refer to the last portion of the small colon. However, if it is spelt with an 'i' to form the word *ilium*, we refer to one of the bones of the pelvic girdle, or coxa.

When we learn, we should learn to learn and focus on letters and words. You will find that certain sequences of letters are often repeated to describe or name structures that may have a common meaning – for example, consider the sequence of the letters **t r a p e z**. These letters in this sequence form many names of anatomical structures, such as the *trapezius* – a muscle in the upper shoulder region – and the *trapezium* and *trapezoid*, which are carpal bones. A trapezoid or trapezium (not in the anatomical sense) is a four-sided shape, which often describes the shape of the anatomical structure, hence the clue in the name.

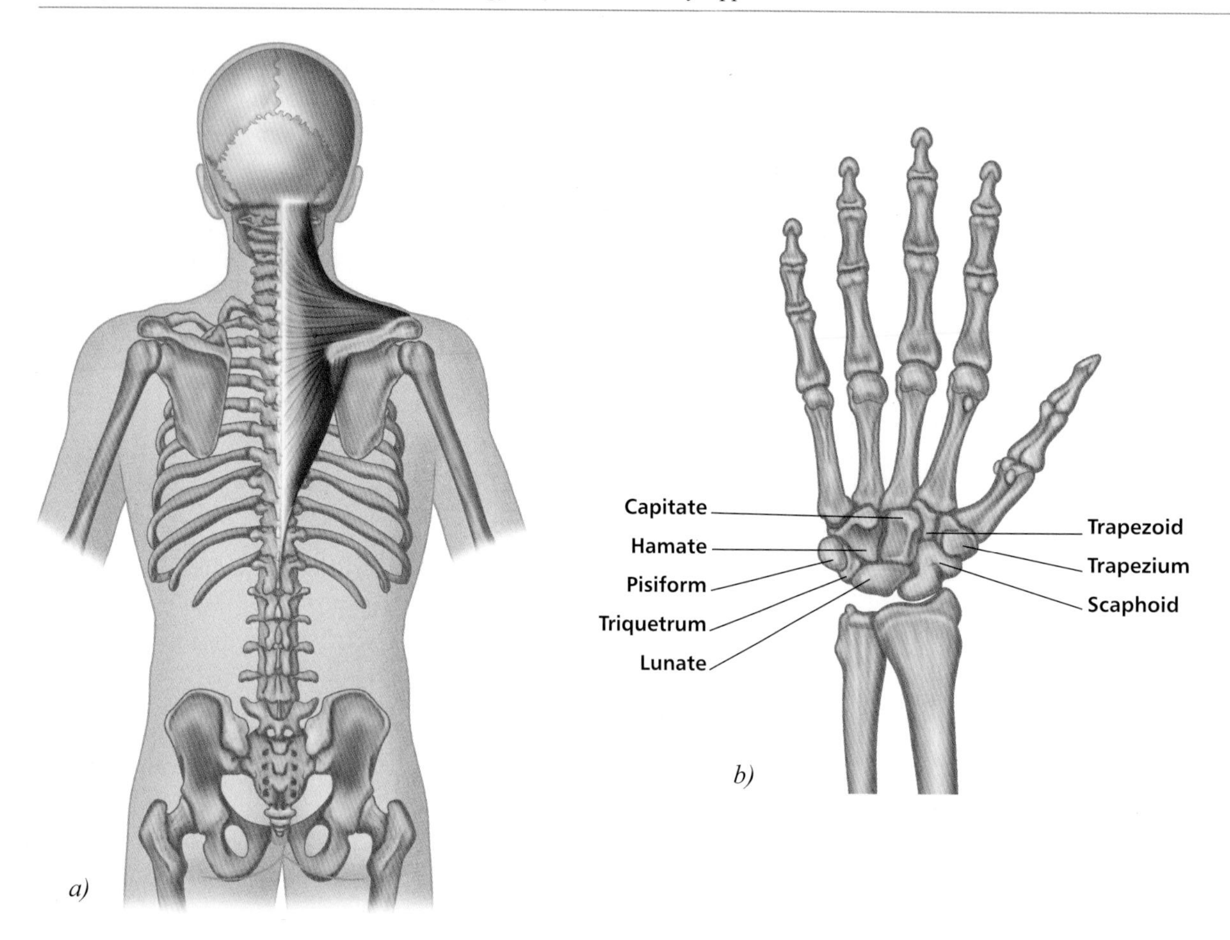

Figure 2.4: a) The trapezius muscle, and, b) trapezium and trapezoid bones in the wrist.

What Is it that We Need to Learn and Learn Well?

It is important whilst learning anatomy to expand your vocabulary with mnemonics by associating funny anecdotes to the term or word you are learning. This helps to secure the word you are learning to your memory more effectively. When learning, it is important to welcome mistakes and learn from them. One of the best things you can do in the initial stages is not to try to get everything perfect, but to embrace that learning environment and the mistakes it may bring.

Finally, think carefully about moving from conversation to mastery. Spend time speaking anatomy with others, practising the words in context, using words to describe body processes and movement. Instead of saying, 'Please bend your forearm,' you may wish to apply your new-found knowledge of anatomy by explaining what bending means anatomically, i.e. 'Please flex your forearm in the sagittal plane.' Your friends may just think you have landed from another planet. At least you have the power to educate them.

Tell Me a Story

Storytelling, or personal narratives, is a powerful method to engage subject matter and apply content to practice. Language development often relies on our ability to collate information and teach ourselves the lessons of the human body. I recall a student who had creatively invented a range of characters which he used in stories to learn the different body systems. The one that stood out most was the journey through the human heart in the form of a rollercoaster ride.

SupVC, InfrVC and CorSi were best of friends. As friends they always stayed together and enjoyed each other's company. One day the friends visited Human Park, a new adventure playground with incredible rides and experiences. The main attraction was 'Have a Heart' – a journey through the human heart. To ensure that each adventure was unique, the three friends had to split up and take separate carriages for the ride. SupVC wanted to start the journey from the head, whilst InfrVC fancied a journey from the feet upwards. CorSi was fascinated with the heart itself and decided to take a carriage from within the heart. The ride began and all three friends entered the right atrium via their respective carriages: SupVC from the Superior Vena Cava, InfrVC from the Inferior Vena Cava, whilst CorSi slid in through the Coronary Sinus. As friends they could never stay apart for long, so they continued their journey together as they dropped through the tricuspid valve into the right ventricle. All three friends marvelled at the chordae tendineae and wanted to tug on the heartstrings, but there was no time as they catapulted past the pulmonary semilunar valves and headed via the pulmonary artery towards the lungs. The friends could not help but feel a little blue as they craved oxygen. Once in the lungs their carriages had time to exchange gases as they squeezed through the thin walls of alveoli and returned to the left atrium via four pulmonary veins. The pressure was mounting as they blasted through the mitral valve into the thick-walled left ventricle, before surging through the aortic semilunar valve into the aorta, to begin their systemic journey through the body and back to the heart. The three friends were excited to meet their exchange friends as they offloaded their carriages of oxygen to the tissues and cells.

The above story illustrates the creative process and content-specific learning that one requires to appreciate the dynamics of the anatomical and physiological processes of the heart. Through stories we can learn to dispel our fears and extract meaning from content. Stories are useful once we have mastered the basic language skills. It is more important to consider putting content into our own words and relating learning to significant experiences and activities.

The human body is a living story – one which has been told many times, in different ways with varying degrees of detail. The story of the living body can begin and end where you like; what is fundamental is that the story of the body parts is a significant component of the whole and needs to be studied and understood as such. The cell, for example, is a microcosm of the entire organism. As you study each system in the body, consider what the main storyline would be:

- What characters would be useful in enacting the process?
- How would you tell the story to a beginner, as opposed to someone with experience, allowing understanding of the systematic organisation of the body?
- Which words or concepts would you want to emphasise in the story?
- Would your story have a twist?
- Which aspects of the story would the reader want to reread?

These questions could assist the learner by directing the learning content and developing language-specific skills.

The Mystery and Marvel of Speaking a Unique Language

When you learn to speak anatomy, you join a unique group of people who are able to appreciate the subtleties within the language. When I learned to speak anatomy, I was so excited that I wanted to share my new-found love for this language with everyone I met. I vividly recall going out for dinner and deciding to impress the waiter with my supposedly vast knowledge of anatomy. I proceeded to order a plate of intercostal muscles (lamb ribs). The waiter looked perplexed and went to place the order. He returned to our table, to report that the restaurant had only local coastal mussels and that inter coastal mussels would be incredibly expensive at this time of year. He politely asked whether I wished to change my order. I then decided to speak English and told him what I wanted. His response was, 'Why didn't you just say that in the first instance?' Humour aside, the message is clear that not everyone fully understands anatomical language and the language could be lost in translation. It is important to identify colleagues who can help you learn and encourage fun whilst you learn.

Lend Me Your Ears – The Art of Listening

> *The most basic and powerful way to connect to another person is to listen. Just listen. Perhaps the most important thing we ever give each other is our attention.*
>
> Rachel Naomi Remen

When we learn, how often do we remove distraction and listen to ourselves and/or the teacher? Listening is a complex skill – one which requires focused attention. Hearing and listening are two separate skills. Hearing is simply noise, whilst listening requires active engagement and extraction of key messages. Learning a language requires us to tune in and use our senses to make sense of what it is we are learning and attempting to master. A good listener pays attention not only to what is being said, but often what is omitted or not said. When we truly listen to words and focus on their meaning, we begin to construct meaning and are able to consider how words become language. Below are a few suggestions to encourage active listening:

1. *Stop talking* – Mark Twain wrote, 'If we were supposed to talk more than we listen, we would have two tongues and one ear.' We need to learn to listen without interruption and with intent.

2. *Prepare to listen and learn* – it is important to adopt a relaxed approached to learning and listening. You need to feel comfortable about accepting the read word and its meaning.

3. *Remove distractions* – when we listen we need to focus on what is being said. This involves actively removing distractions.

4. *Be patient* – remember, listening is a complex skill that may require adjustment and time to master. Listen slowly and progressively to the way in which anatomy as a language is used and spoken.

5. *Listen for ideas, not just words* – consider what is being communicated. How can this message be translated or transformed into a storyline that you could relate to others? What specific skill is needed to link ideas together and construct new knowledge?

In summary and conclusion, I would like to leave you with a few thoughts about learning and teaching anatomy as a language:

1. *Know why you are doing it* – it is important that learners and teachers are clear about the reasons for teaching and learning anatomy and the techniques that will be used to foster learning and application.

2. *Dive in* – practise anatomy daily; be excited about the fact that you are learning a new language; tell people that you are learning to speak anatomy.

3. *Find a friend to learn with you* – it is useful to learn with someone, someone who understands and appreciates what it is you are trying to achieve.

4. *Keep it relevant* – when you learn anatomy there may be many distractors; focus on one theme at a time and practise that

5. *Have fun* – think of fun, creative and innovative methods to bring your language to life. You may wish to create characters to help you learn names and positions of structures, e.g. Mr Femur is your typical hippy and does lunch with Ace Tabulum – my word, when the two of them meet, do they move.

6. *Act like a child* – don't be scared to explore the language, make mistakes and learn from them. Don't be scared to admit, 'I haven't learned that yet,' as opposed to 'I can't' or 'I don't'.

7. *Leave your comfort zone* – willingness to make mistakes means that you could be vulnerable and easily embarrassed. This can be scary but, equally, transformative. Next time you visit your doctor, you may wish to explain your issue in a language you both understand.

8. *Listen* – when we learn a language we must learn to listen before we speak. Anatomy as a language may sound strange the first time you hear it being spoken, but the more you expose yourself to it, the more familiar it becomes, and the better able you will be to make connections between words and terms. How often have you picked up a word from another language and thought about it and then realised – 'Aha! So that's where the word originated …'

9. *Watch people talk or teach anatomy* – this will encourage you to understand how words are enunciated and described. You may even wish to read medical reports to identify anatomical terms.

10. *Use self-talk* – talk to yourself by revising and refreshing anatomical terms. Build a vocabulary and focus on developing meaning for the words you use. When you sit on a bus or train, describe the person opposite you in anatomical language – just be careful to control your laughter!

Final Thoughts

It is not solely about speaking anatomy, but more about how comfortable you feel within the process.

This chapter has simply provided a framework for positioning the learning of anatomy as a language within the reach of the learner and teacher.

3 Do Anatomy

I profess both to learn and to teach anatomy, not from books but from dissections; not from positions of philosophers but from the fabric of nature.

William Harvey

As infants and children, we learn from our experiences; we copy our parents and carers and respond to our environment in ways partly that we are preprogrammed to do, and partly that we have learned. We have our ability to smile winningly at our parents, or blow raspberries at Uncle Tim, because we have learned to do so via a variety and combination of techniques. We didn't have formal lessons in blowing raspberries; no one made us take notes or give in homework assignments. Instead, we used our innate reflexes and responded to those around us.

This childlike way of learning can be used to help us learn and teach anatomy and physiology. Although childlike, these methods are complex and involve four main processes of multisensory learning which have the acronym 'MARS' (American Red Cross):

- Motivation
- Association
- Repetition
- Sense

We will look at each of these processes in turn, then unpack ways that the theory can be turned into actions. This enables practical approaches for getting the learning across to learners in a variety of ways, and practical methods for learners to take control of their learning, own the information, and allow anatomy and physiology to make sense.

Motivation

We are all familiar with the need to clean out the recycling bins rather than complete an anatomy essay! Learners need to have the internal drive and motivation to dedicate time and effort to learning anatomy. Procrastination is somehow far more appealing than knuckling down to the work. Self-motivation can be achieved by setting deadlines and targets. Using a timer can motivate a learner to concentrate for a set amount of time – or perhaps to enjoy procrastination for only a set amount of time before cracking on with the task.

From a teaching perspective, motivation can be helped by making the learning interesting. Those recycling bins will always be more fun than a dry, dusty lecture and an even duller homework assignment. Teaching with the VARK learning styles (visual, auditory, read/write, kinaesthetic) in mind and using a variety of methods will hold learners' attention and concentration (see Appendix for VARK questionnaire). However, learners come from a variety of backgrounds and cultures, so are motivated by different things, and maintaining motivation for everyone is nigh on impossible; after all, we all have good days and bad days. The best approach in teaching anatomy is to keep the lesson and task varied and the pace of class varied too.

One of the best forms of motivation in a classroom setting is FUN! Learning with a smile on your face is not a chore. Creative and supportive teaching and learning will form a welcoming environment where learning is a given because all barriers to learning have been removed. Inject a sense of humour to that mix and you have an unbeatable formula! Laughter is infectious; if the teacher can create a learning environment where smiles, chuckles or downright laughter is commonplace, the learners will become involved too, and the process will continue amongst the learners, both in and out of the classroom. This does not mean that all teachers need to go to drama school in order to brush up their 'stand-up' routines. No disco lights, dry ice or loud music needed. Instead it means that teachers have to manage their own emotions and mental states, so that any angst in their own lives is left well outside the classroom, and they bring in a sense of fun – looking for positivity and creativity in all of the subjects being taught.

Association

All our knowledge has its origin in our perceptions.

Leonardo Da Vinci

Finding the links between ideas and how things fit together and affect each other is one of the biggest keys to learning. In order to make sense of the world around us, we learn to look at cause and effect, and the associations between objects and events. The more links that we see, the better our lives make sense. Our brains have amazing capability to do just this.

Just as we learned to recognise that Uncle Tim really enjoyed our raspberry blowing when we were a few months old, we will be able to better remember the connections between body systems and the relevance of the minutiae of detail when things are seen in context.

Jill Bolte Taylor, a neuroscientist, wrote about her experiences of recovery in her book *My Stroke of Insight*, just after she suffered a catastrophic brain haemorrhage, damaging her information-processing centres:

I took the sounds of the key words and repeated them over and over again in my brain so that I would not forget what they sounded like. Then I would go on a process of exploration to identify meaning that matched the sound of these words. President, what is a President? What does that mean? Once I found a concept (picture) of what a President was then I moved on to the sound United States. United States, United States, what is a United States? What does that mean? Once I found the file for United States, again, it was a picture in my mind. Then I had to put together the two images – that of a President and that of the United States. But my doctor was not asking me anything that was really about the United States or about a President. He was asking me to identify a specific man, and that was a completely different file. Because my brain could not get from 'President' and 'United States' to 'Bill Clinton', I gave up – but only after hours of probing and mental gymnastics.

Jill Bolte Taylor, 2008, *My Stroke of Insight*, p. 77

So, extrapolating from Jill Bolte Taylor's experience, many learners attempt to memorise isolated anatomical terms and facts then wonder why they do not recall much information at the examination. Interpretation of the facts is required when answering examination questions and when applying the anatomy and physiology information in the workplace. Without connections and associations the information is pointless, and nothing more than trivia.

The reductionist approach of looking at individual body systems or body parts is a fine way to specialise, but in order to learn and memorise, finding the links between the systems is imperative. Contextualising the information, such as looking at the medical diseases of a body system or body part, makes a more interesting story, and stories are more memorable than bald facts.

Tell me and I forget. Teach me and I remember. Involve me and I learn.

Benjamin Franklin

Repetition

When learning a skill such as playing the piano, it is pretty obvious that repetition in the form of practice and performing scales and arpeggios is pretty rudimentary. We are unable to perform a concerto at the Royal Albert Hall without years of intensive practice. So why, then, is it that we forget about repetition of anatomy knowledge and skills? Learners must train their brains to think in an anatomical way, and to recall facts and figures on demand. All of this takes practice. Just think, how many times does a toddler fall down before he or she is able to walk steadily?

Anatomy and physiology are often perceived as theoretical subjects, so learners assume that repetition involves writing out lists and tables, over and over again. Actually, they can be seen as practically based subjects, where the information can be contextualised, so that the learner's own body becomes a tool for learning. After all, the learner's own body is the only 'cheat' that can be brought into the examination, so it makes sense to use the body's actions to memorise anatomical information. Repetition would involve repeatedly moving the structure in different directions to work out ranges and directions of motion for each joint, or to locate structures and point them out over and over again.

Figure 3.1: Examination students using their own bodies to help answer the questions.

Anatomy and physiology courses may be taught at different levels, from the simplest information learned at school, then further developed at subsequent course attendance. By revisiting the information from previous courses, the learner can see how the foundations in anatomy are laid down, and often what was perceived as hard and nearly impossible to learn, years ago, becomes much more manageable when reviewed months later.

In order to fully remember a subject, the information needs to move from the short-term memory to long term. The first time a subject is introduced, the information sits in the short-term memory; if it is not used, pulled apart or applied, but just left alone to gather dust, the memory drops away fairly quickly. Active learning and interaction with the retained information, together with repetition of the information, will cause a large portion of the information to be converted to long-term memory. Again, the same process will occur – if not utilised in any way at all, the memory will drop away again, but usually the time it takes to do so will be slightly longer. This process will continue, and each time, the length of time the information is retained will be longer and longer. It is important to remember that each learner will have his or her own individual set of optimal timings for repetition, revisiting and revising in order to maximise this process.

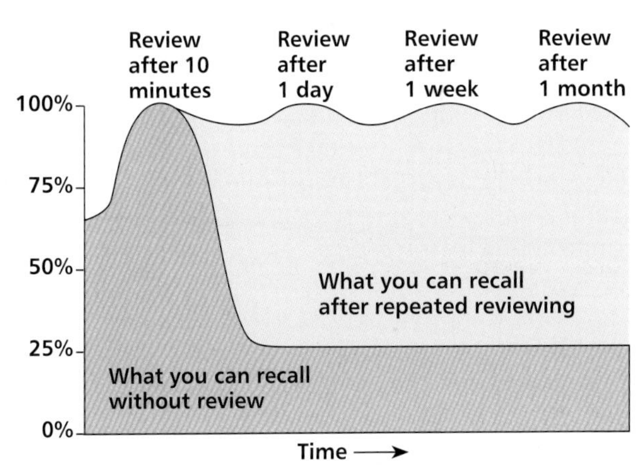

Figure 3.2: Graph showing learning/revision cycle.

Sense

I hear and I forget. I see and I remember. I do and I understand.

Confucius

Making sense of anatomy and physiology requires active learning, using our own senses. We will be discussing the visual nature of anatomy in more detail in Chapter 4 – See Anatomy.

Children learn through their senses and react according to their interpretation of their environment. Using the senses of vision, hearing and touch, we, as adults, can hone our teaching and learning by using all the senses according to the VARK (visual, auditory, read/write, kinaesthetic) system.

Students retain:
10 percent of what they read,
26 percent of what they hear,
30 percent of what they see,
50 percent of what they see and hear,
70 percent of what they say, and
90 percent of what they say as they do something

Stice (1987)

The pie chart in Figure 3.3 shows the distribution of the VARK learning preferences. It can be seen that only 3% have purely visual learning preference, 7% have purely aural learning preference, 12% have purely read/write preference and 15% had purely kinaesthetic learning preference. The vast majority of learners have a combination of learning styles. It follows that, as teachers, we need to make sure that the classroom and textbook delivery uses all of these senses, and that learners involve as many senses as possible in order to learn, understand and make true sense of the information.

Learners can assess their own learning preferences by completing the VARK questionnaire in the appendix of this book. As we go through each chapter, each of the senses and learning styles will be examined in more detail and practical suggestions will be given for different learning and teaching strategies to suit all learners' needs.

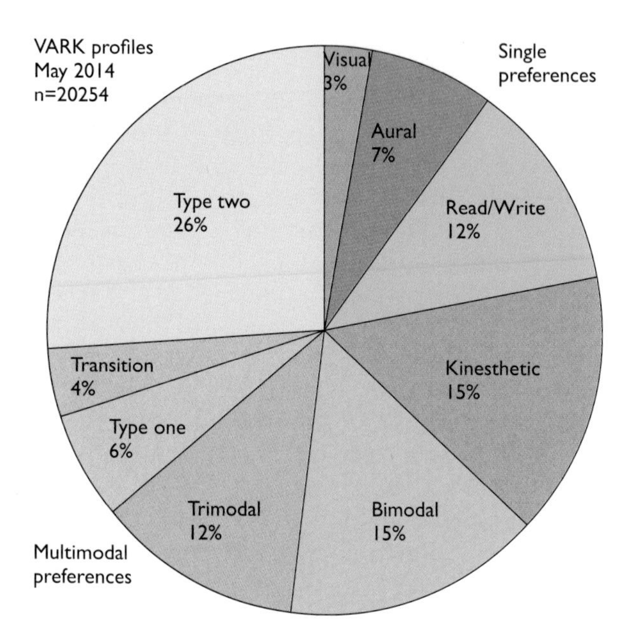

Figure 3.3: Distribution of learning preferences using VARK. From VARK Learn Ltd (2015).

Visual Learning

The visual learner will 'see' your point, and will prefer to look at pictures, PowerPoint presentations or diagrams rather than reading screeds of text or listening to a lecture or podcast. Using the following tools to teach and to learn anatomy and physiology will be most useful.

PowerPoint

The ubiquitous PowerPoint presentation (or one made using similar presentation software) is the bane of many learner's lives. The phrase 'death by PowerPoint' is as commonplace in the classroom as the desk and chair. Many teachers will use PowerPoint as a way to get as much information on to a slide as possible, in order to teach from the slide. This may create a detailed and complicated image to those learning the subject for the first time. Many of us will have been bored rigid by teachers who simply read from a visually busy slide, rather than using key words and phrases displayed in a clear format.

Tips for a PowerPoint Presentation	
Keep it simple	Plenty of white space means less visual clutter Make the information the star of the show, not your IT skills Simpler visuals mean easier-to-remember information
Limit text	Keep to bullet points Narrative should be spoken, not on the slides Provide a detailed handout afterwards to support the slides
Limit animations	Fancy 'transitions' are tedious to watch Use a simple 'wipe left to right' for best impact Avoid too many variations of transition (two or three types maximum)
Keep to a theme	Work out a style guide to match your class, and stick to it Varying themes in a single presentation, due to poor cutting and pasting, is distracting
Use colour appropriately	Choose background colours depending on classroom lighting o Dark background for large, dark rooms o White background for light rooms
Use quality images	Use high-resolution images to avoid degradation of resolution Use real photographs rather than line diagrams as much as possible, to keep things in context Only use high-quality line diagrams with clear labelling Avoid cartoon 'clip art' unless it actually matches your point
Use the right font	Avoid serif fonts o Too visually fussy when close up o Serifs sometimes get lost in poor resolution Choose sans-serif fonts like Gill Sans o Appears professional, friendly and more conversational Choose correct font size for room size Avoid using more than one font set
Use video or audio	Embed video or audio clips o Adds an auditory layer of learning to the visual o Adds a visual layer of learning to the auditory
Limit cheesy effects	Builds irritation in the learner Causes learner to lose motivation

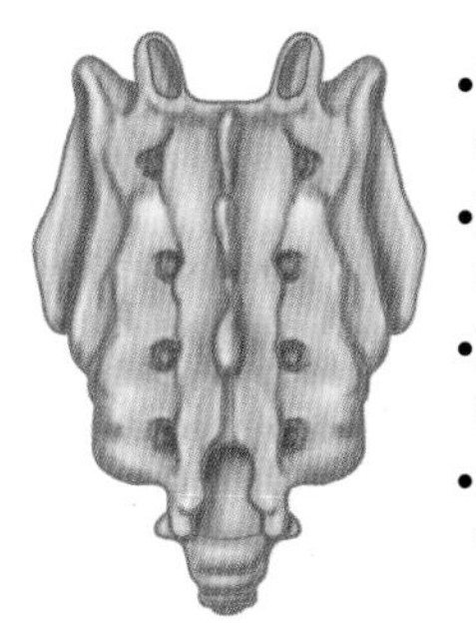

- The sacrum is a large, triangular bone
- Positioned at the base of the spine
- Formed by five fused sacral vertebrae
- Connects with last lumbar vertebra and coccyx

The sacrum (sacred bone) is a large triangular bone located at the base of the lumbar spine and forms the back part of the pelvic cavity. The sacrum starts out from birth as five individual bones before starting to fuse between the ages of 16 and 18; the sacrum is considered to have fully fused into a single bone by the time you have reached 34 years of age. Considerable differences in the shape of the sacrum between individuals, as well as structural differences between the left and right sides, are well documented. The connection of the sacrum to the ilium forms the sacroiliac joint.

Figure 3.4: Examples of: a) a good, clear slide, b) a poor, cluttered slide.

Labelling Diagrams

Much of anatomy learning involves the name and location of a structure. Labelling preproduced diagrams is a useful way of working out name and location, and the repetition of this will reinforce learning. The best diagrams are ones which are clear and 'friendly' to the eye, where the arrows pointing to the area in question are clear and do not overlap each other. Simple diagrams will be easier to visually recall than messy diagrams.

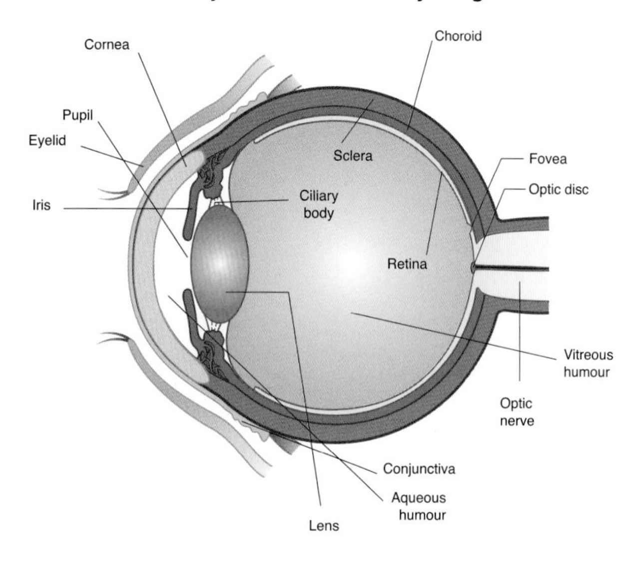

Figure 3.5: Example of a poor diagram with overlapping arrows.

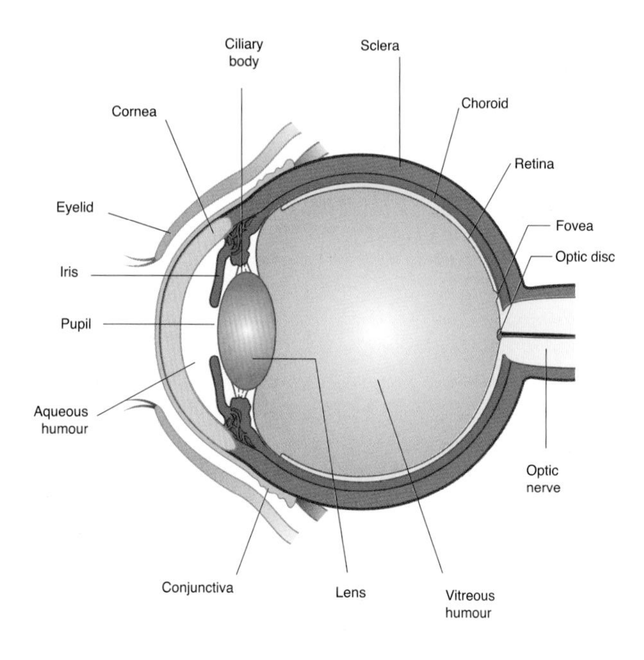

Figure 3.6: Example of a clear diagram of the eye.

When testing oneself as revision, in the first instance, diagrams can be devoid of all arrows, and a list of items to be located can be used.

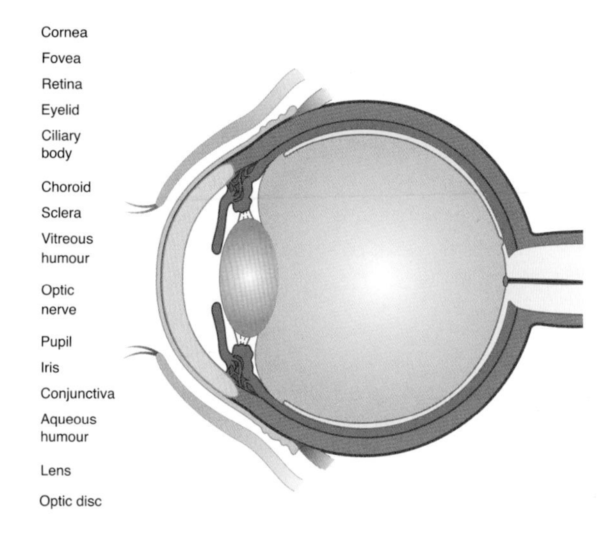

Figure 3.7: Example of a diagram with no arrows and a list of labels.

A labelled diagram is also a good starting point for a mind map. The labelling can be expanded to create a visually memorable summary of information that is portable, so easy to carry around and glance at regularly to reinforce learning.

Palpation and Drawing on Skin

Learners enjoy this one. The one 'cheat' that can be brought into the examination is the learner's own body, so what better way to learn the location of structures than by utilising the body and reinforcing learning by actively drawing on these structures? Making sure that all traces of paint and ink are removed before the examination, of course.

Use diagrams and images in textbooks to locate key anatomical landmarks. Ensure that full consent is given by all participating students. Palpate the structures by locating the landmarks and learn to interpret the kinaesthetic picture being built. Ensure that there are plenty of wet wipes and make-up-removing cleansers. It is useful to use eyebrow pencils to draw bones, muscles, ligaments and organs directly onto the skin. For longer-lasting drawing, use a washable felt-tipped pen. Even if art was not a strong point at school, organs and structures can be discussed, palpated and located on the skin in a light-hearted and memorable way. The anatomical landmarks and structures may have palpable features, such as the direction of muscle fibres, or the shape of a joint, which are worthy of discussion. The artier students may enjoy painting on the structure in its full detail using body paint obtained from professional make-up suppliers.

If direct drawing on skin is inappropriate or unwanted, marker pens on plain T-shirts work well for abdominal and thoracic organs. Plain pale-toned leggings could be used for drawing on leg muscles, ligaments and bones.

Taking this one stage further, clothing can be purchased that has been preprinted with correctly positioned anatomical structures: socks, T-shirts, dresses and hats can demonstrate skeletal structures and brains – very memorable indeed!

A creative application of this is used by the 'Immaculate Dissection' team from New York. Dr Kathy Dooley and her team teach the rudiments of anatomy by using models exquisitely body-painted by the very talented artist Danny Quirk. Since the painted structures are absolutely lifelike and life-sized, they also move as the model moves. The understanding of the effects of contracting and concentrically loading certain muscles is obvious. Furthermore, the physical connections between adjacent muscles and the importance of the connective tissue and fascia in mobility is crystal clear and unforgettable. The models can be palpated at exactly the right place in order to locate and feel the structures being taught, simply by putting hands directly onto the painted image – perfect for the kinaesthetic and visual learner.

Figure 3.8: Anatomical clothing showing just some of the examples that are available.

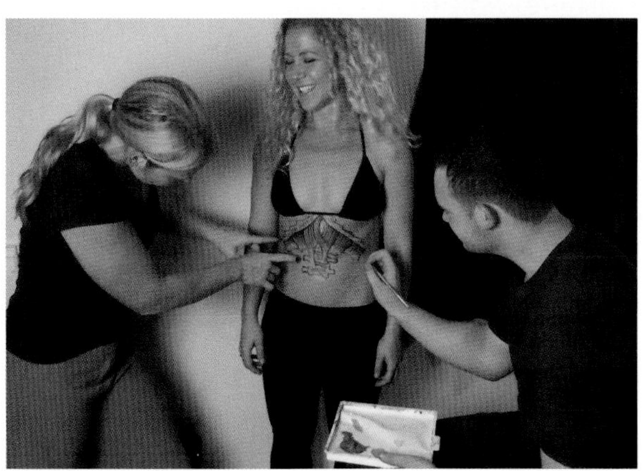

Figure 3.9: Use models who have been body-painted to teach and learn anatomy. (Shown here are the Immaculate Dissection team: Kathy Dooley, Anna Folckomer, and artist Danny Quirk)

Videos

Many textbooks now have related websites and DVDs to accompany the course. Learners will find them an invaluable resource to support the textbook. The video will not include quite as much detail as a textbook, so should not replace the written text, but will complement it and can be watched over and over to revise and embed knowledge.

Some learners may prefer to use distance learning, which often involves watching online seminars or lectures. The drawback for this method is mainly the inability to ask questions at the time of the lecture, although some course providers offer the lecture in the form of a 'webinar', where interactive questioning is encouraged. Provided the online course has sufficient tutor support and mentoring available, this can be an effective way of learning anatomy and physiology and is an accessible option for many who are unable to attend a college or class owing to family or other commitments.

YouTube for Dissection

Going to visit a dissection or prosection is an important part of learning anatomy in university, but is not an option for the many people who are not learning anatomy and physiology in a university or medical school. There are many reliable and respected sites on YouTube which offer videos of prosections and dissections post-mortem.

Visual learners will certainly appreciate seeing the location and depth of tissue of some of the structures which may not be easy to visualise when they are normally depicted in a textbook, often in a stylised form. The more squeamish learners will find it easier to watch a dissection online than in the post-mortem room or dissection laboratory, as it offers a dissociation from emotion, sounds and smells.

Figure 3.10: Dissection of the lower leg: the red cylindrical protein we all know as muscle fibres can be found as muscle islands beneath the skin or on the outer surface of a tendon. (Photograph courtesy of J. Sharkey 2010)

Virtual Anatomy via Computer Software, Apps, etc.

In the absence of watching real human dissection online, there are many apps and websites which offer stylised anatomy lectures, often with interactive features. Examples are included in Chapter 9 – Resource Anatomy.

Anatomy Colouring Books

These were very popular when I started learning anatomy. They are line diagrams of structures that need to be coloured in using designated colours according to a key, in which the name of the structure is often also coloured in with the same coloured pen. Current fashion has seen the advent of adult colouring books as a relaxation pursuit, so students who enjoy such activities may well enjoy completing anatomy colouring.

They seem a great idea in principle, and, as we will see in Chapter 4 – See Anatomy, are effective for most visual learners. Just a gentle warning for visual kinaesthetic learners, who like to 'do' as well as 'see': these types of learners often need the colouring in to be just perfect. They may spend hours neatly colouring in, so the ink doesn't stray over the black outlines, to the detriment of their ability to focus on the structure itself. Hours can be wasted on a pretty diagram, with absolutely no recollection of the names and locations of the structures at all. So colouring books should be used judiciously and only for people who can keep up the focus on the task in hand.

Microscopy

Detailed microscopy is an interesting way of tuning in to the intricate structure of an organ. Learning the shape and layout of cells specific to an organ, or necessary for the function of that organ, may be very useful to a visual learner, but less so for an auditory learner. Encouraging a learner to sketch the cells will reinforce the visual image, and labelling the diagrams with structural and functional information will consolidate things further.

An example of microscopy as a tool could be when learning about the blood. A description of a red blood cell (erythrocyte) as a 'biconcave disc' that deforms to pass through capillaries is not nearly as memorable as looking at pictures of photomicrographs of red blood cells, and seeing just how much they really do look like doughnuts. Likewise, learning about sickle cell anaemia as a genetic condition will be made far more interesting and memorable if the sickle-shaped cells are visualised; imagine those trying to squeeze through the capillaries? What happens when they get stuck?

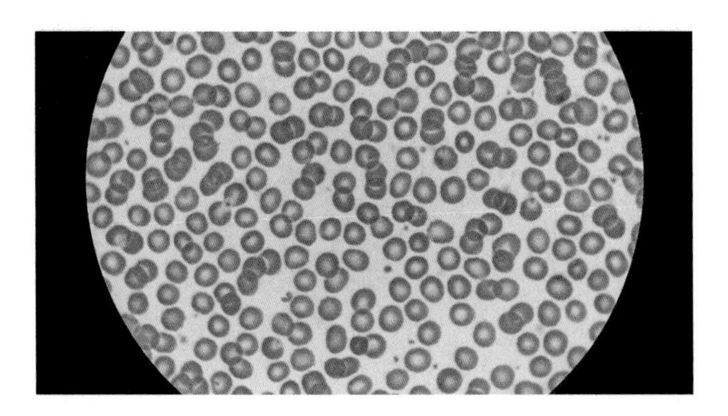

Figure 3.11: A photomicrograph of red blood cells – note their doughnut shape. (Photograph courtesy of shutterstock.com.)

Museum Trips

Search around for anatomical museums. These hold a wealth of pathological and anatomical specimens, and are worth a visit to help visualise the structures. In the United Kingdom, the Royal College of Surgeons has such museums – the Hunterian Museum and the Wellcome Museum of Anatomy and Pathology – as does Guy's and St Thomas' NHS Foundation Trust. Teaching hospitals may have a museum too.

Plasticised cadaver exhibitions such as the Body Worlds exhibition are a superb way of seeing the inside of the body although they are few and far between. Do make an effort to attend if there is one near you.

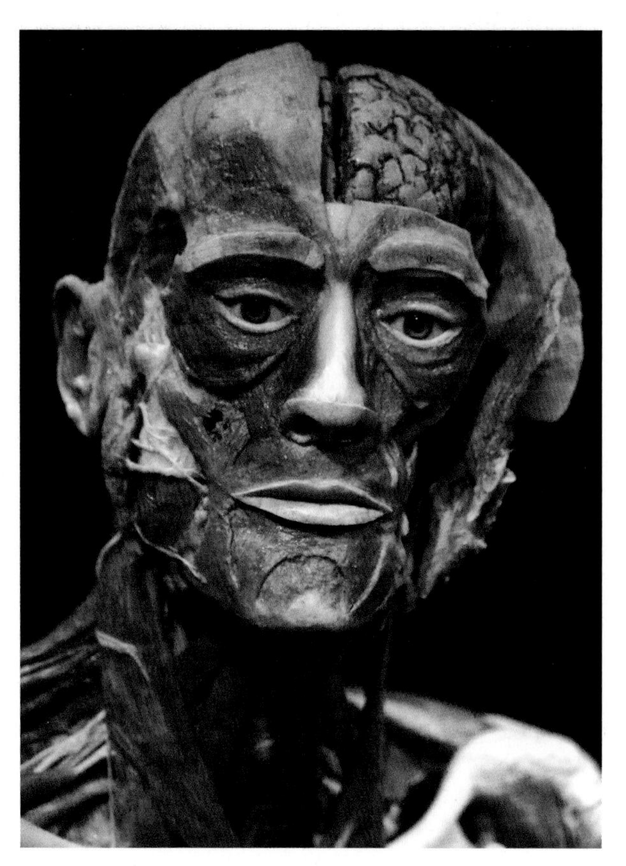

Figure 3.12:A fully plastinated human body in the newly opened "Menschen Museum" (human museum) with plastinated corpses at the Alexanderplatz, Berlin-Mitte. (Photograph courtesy of shutterstock.com.)

Auditory Learning

The auditory learner will 'hear' your point, and will argue points by saying 'Listen up' or 'Listen here!' Often the auditory learner needs visual input as well to reinforce the auditory input.

Standard 'Listening to the Lecture'

Sitting in class and just listening to a lecturer rarely works for the majority of learners. The auditory learner would be in the minority – but in order to succeed in learning anything from listening, needs an uninterrupted 'learnscape'; no irritating clicking of pens in the background, no sounds of other people typing or texting, no hum from the projector or air-conditioning unit. Auditory learners are easily distracted by sound!

If you are an auditory learner, sit near the front so visual disturbances from others don't put you off. Sit away from the noisier classmates and integrate the visual information presented by taking notes. Reading your written notes with an internal voice will reinforce and repeat the auditory input.

Recording Lessons

Using a voice-recording app to dictate course notes on to an MP3 player or mobile phone will be invaluable to the auditory learner. Playing these back to listen to whilst walking is the best way of revisiting the class.

Additionally, the voice recorder can be used to record a lecture. Obviously, permission must be granted by the lecturer in order to digitally record the class.

Podcasts

If auditory learning works for you, go online and shop around for podcasts of your chosen topic. These can be listened to at convenient times of the day: walking the dog, driving, cooking or at bedtime.

Audio Textbooks

Many good anatomy and physiology textbooks have an audio file on their associated website. Downloading the audio file enables one to listen to the textbook. Drawbacks in this method may be that the voice which is reading the text is found to be unpleasant or has an unusual accent and this will break concentration. If this is a problem, record the text yourself and so you hear your own voice. An advantage in this method of learning is that the recording can be played while walking or driving.

Poems

As we saw in Chapter 2 – Speak Anatomy, anatomical language has a rhythm of its own. An auditory learner will love to make up poems and sayings that have a regular rhythm, thus making them more memorable.

Example:

There was once a weak one-sided psoas

That caused a spine S-bend and sore ass,

Scoliosis abound

The shoulders were round

That tenderloin muscle called psoas

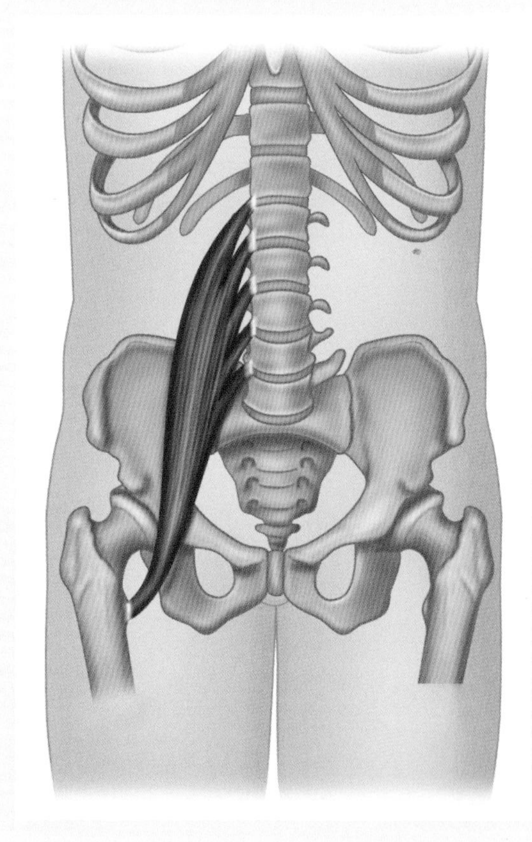

Figure 3.13: Illustration of the psoas.

Example:

The 12 Cranial Nerves

(Sung to the tune of *The 12 Days of Christmas*)

On the first cranial nerve

My true love sent to me

My sense olfactory

On the second cranial nerve

Two optic nerves …

Three oc'lomotor …

Four sweet trochlear …

TRIGEMINAL

Six abducens …

Seven facial nerves …

Eight auditory …

Glossopharyngeal …

Ten vagus nerve …

Spinal accessory …

Twelve hypoglossal …

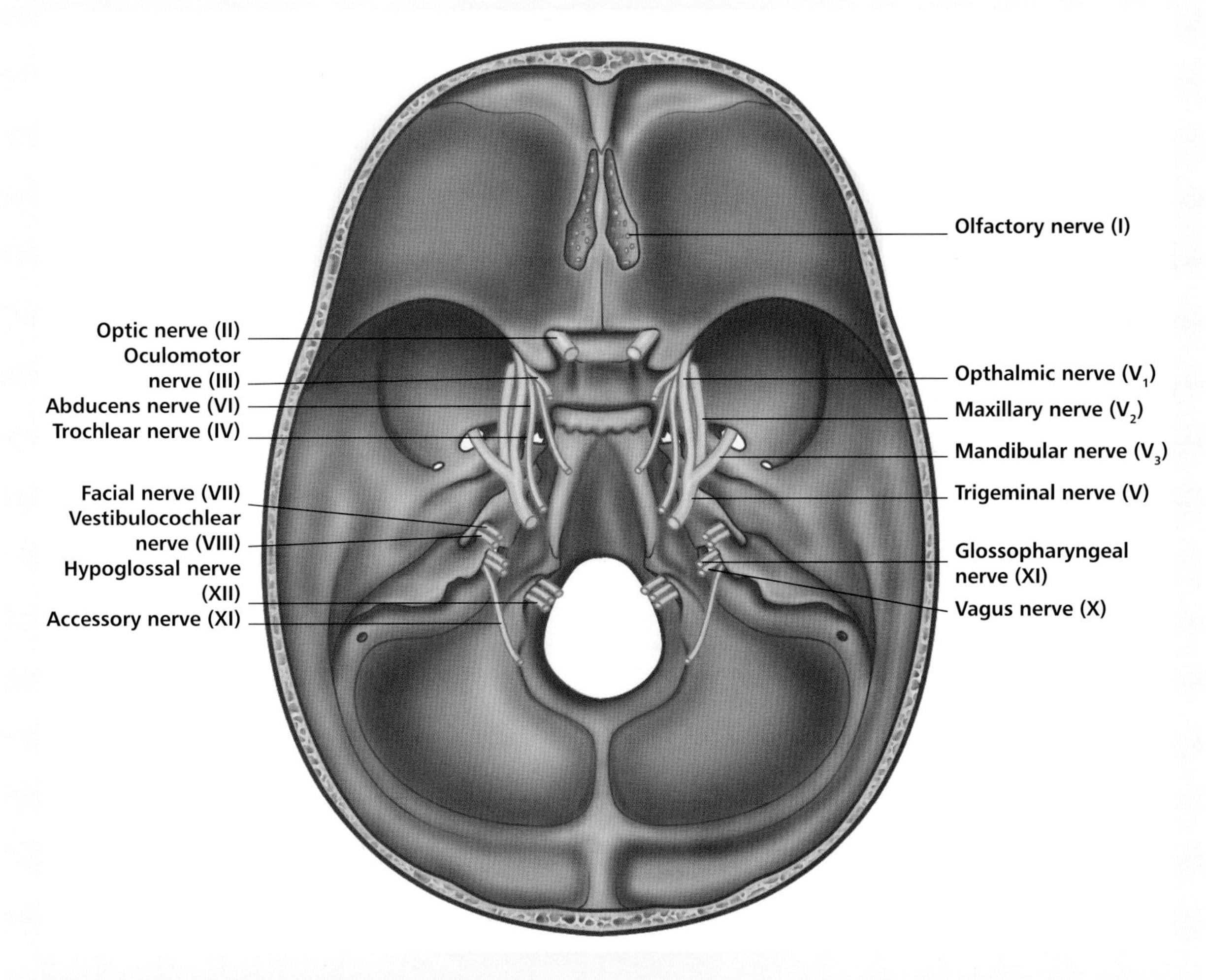

Figure 3.14: Illustration of the cranial nerves.

Example:

Bone/Skeleton Functions – SADPAM

Shape of body and support other tissues
Attachments for muscles and ligaments
Development of blood cells
Protection of vital organs
Allows movement
Mineral store and lipids

Or

Shape of body	Special
Protection	People
Mineral storage	Mend
Lipids storage	Limbs
Movement	Medically
Blood cell production	Better

Facial Bones

Mandible	My
Maxilla	Mother
Nasal	Never
Lacrimal	Liked
Zygomatic	Zipping
Palatine	Pants
Vomer	Very
Turbinate	Tightly

Cranial Bones

Parietal	Pants
Sphenoid	Should
Frontal	Fit
Temporal	Tightly
Occipital	Over
Ethmoid	Everything

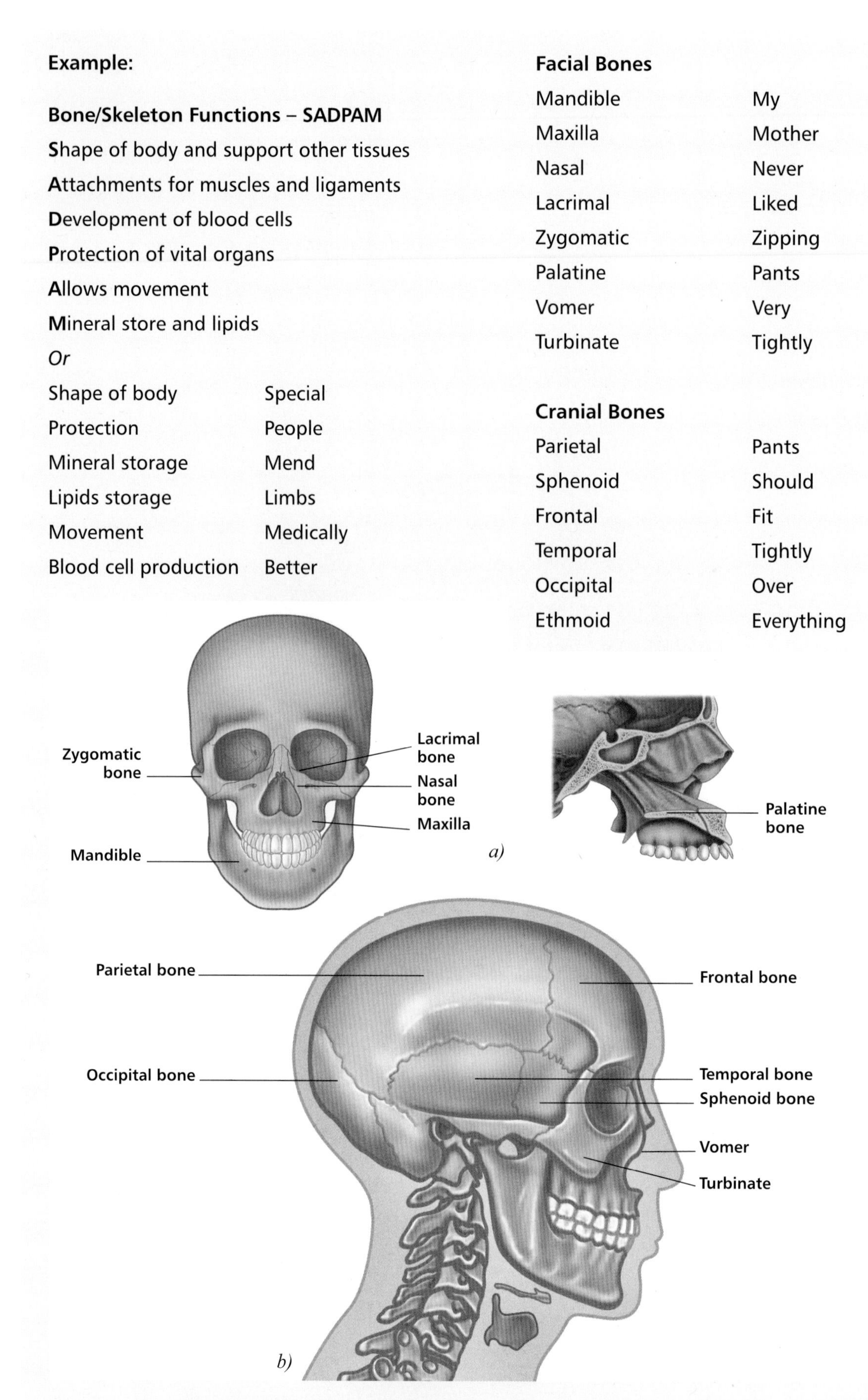

Figure 3.15: a) Facial bones, and, b) cranial bones.

Songs

Combining memorable poetry with music is a winner for the auditory learner; either writing your own, or listening to that written by others. Combining the song with actions will embed the information for all learning styles. My own students have NEVER forgotten the name of the fifth cranial nerve!

Mnemonics

It is a fact that anatomical and physiological information requires lists. Fact! So make the lists memorable by using mnemonics, and spice them up by being cheeky!

Read/Write/Draw Learning

The read/write learning style is one which most learners think they have – although that may not be the case! It is the preferred and usual method for studying for most people, even though for many, it hasn't necessarily worked for them before and is definitely not the most effective method. Reading and writing uses both visual and auditory learning, as the internal voice reads the written word.

My advice for those who wish to use this method is to look back at previous learning experiences; did this work for you before? If you struggled to remember the information for an examination in the past, you are unlikely to actually learn best with this method, however comfortable it may feel. So, step outside of your comfort zone and use some of the other learning techniques as well as read/write. In particular, do not rely on *just* reading!

Note Taking Whilst Listening to the Lecture

Not the best method for learning when used in isolation, but certainly a good way of recording the lecture. Improve by annotating a PowerPoint slide or diagrams that have been prepared in advance. Teachers could provide these in advance, as these will have already prepared.

Internet Searches

Many people spend hours a day reading the results of internet searches. One search leads to another and before you know it, you have wasted the best part of an afternoon and managed to learn about talking cats, instead of the structure of a lumbar vertebra! Be disciplined in your internet searching, and save the searches to folders or to appropriate bookmarks so you are able to find them again.

Standard Essays and Homework Questions

Conventional homework requires a learner to read and research the anatomy and physiology using textbooks, presentations or classroom notes, and assimilate this into their own words in order to answer the homework essay or question. Using these textbooks needs thought and technique. It is only too easy to glance at the written word and to be unable to convert those words into your own, or to turn the paragraph around to get the salient points in order to actually answer the question, rather than regurgitating everything you know about the topic.

The best way of using the textbooks is to *activate* your learning. This could be in the form of annotating your textbook; horror of horrors for many people! Alternatively, use sticky notes that can be plastered all over the textbooks with your key words, phrases and diagrams stuck on rather than scribbled directly to the book – aesthetically pleasing to both 'perfect book' lovers and to stationery lovers alike, particularly when the sticky notes are colour coded!

Planning the essays requires technique too. The use of mind maps, spider diagrams, flow charts or lists helps to structure the answer into broad topics which can be elaborated upon to form the full answer. This will also allow your own personal style to shine through rather than using words directly lifted from the textbook.

Good homework and essay questions allow the learner to consolidate the learning from previous lectures as well as to prepare for upcoming classes. Familiarising yourself with unfamiliar terminology and complicated processes is very worthwhile. Learning something completely for the first time is always tough on the brain – learning something the second time around is always easier, so put in the preparation in advance.

Where possible, include labelled and annotated diagrams in the answers (see Figure 3.16). It has been said that a picture speaks a thousand words, and certainly the brain organisation used to source the picture, copy it, and thoughtfully label and annotate it will consolidate the information.

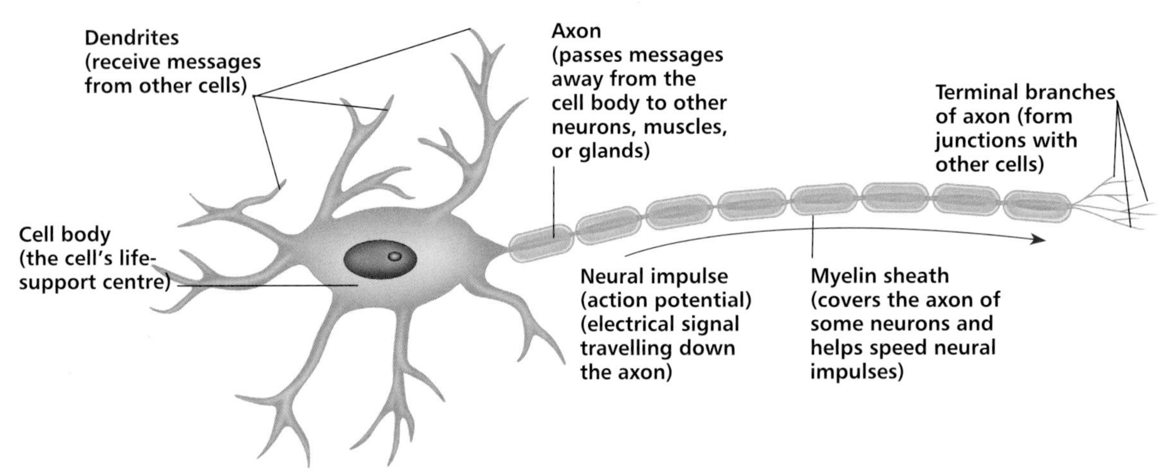

Figure 3.16: Picture of an annotated diagram of a neuron.

Textbooks With Student Workbooks

Choose the course textbooks carefully. This applies to both learners and teachers of anatomy and physiology. Often students scurry to the nearest bookshop, or are addicted to scrolling through pages of beautiful books online in order to purchase the perfect book. Take heed and choose wisely!

Firstly, the textbook should be representative of the level of anatomy and physiology being assessed. Too often, a recommended textbook is of too high a level of learning, so a student is instantly repelled as they cannot relate to its language; they immediately feel alienated as they do not understand what appears to be basic information. Instead, the learner will go into a downward spiral – 'It's too hard for me,' 'I can't possibly learn all this' and 'I will never understand all this in a million years' have all been heard over the years.

Teachers must look at the intended learning outcomes written at the start of the anatomy and physiology courses and work out the level that the final examination is intending to set, and then find a textbook that offers information very slightly above that level of learning. By setting the reading level slightly above the level being examined allows for extension material, and creativity and interest in the most able and driven students, whilst being manageable in language, layout and content for the average learner.

Secondly, the book should be appealing to the eye. Having an aesthetically pleasing typeface makes for an easier read, so pick a book that has a clear typeface at a reasonable font size. Many a student has been scared away from reading if the typeface is simply too small. The diagrams and tables should be visually attention-grabbing. Look for the use of colour. Two-tone textbooks are dull – yet migraine-inducing rainbow books can be visually overloading. Aim for what suits your eye.

Thirdly, the layout of the book should be clear and the information should be relatively easy to find: contents pages, index page, glossary and references should all be easy to locate. The use of bold type where there are glossary entries is very useful, particularly for beginners to anatomy, who are learning a new language as well as learning new information.

Finally, the book must be portable. This may seem really flippant but it applies to all textbooks, not just those of anatomy. Textbooks are meant to be read and reread. Leaving them at home or on a shelf because they weigh a ton is a pointless waste of money. Have constant access to your textbook. Choose either a soft-backed book that feels the right weight for you, or choose one that has online access, either as a downloadable e-book, or with resources available online or via an app on your mobile device. Some anatomy textbooks are also available as audio books, making them suitable to listen to whilst walking or running, or even as background whilst driving or working.

Teachers: be mindful of choosing books that only match your own teaching style. Instead, be very aware of learning styles, and choose a textbook which will meet the needs of all of the learners as well as matching the level and style of the class.

Interactive Reading and Completing the Workbook

Some of the best anatomy and physiology textbooks have accompanying student workbooks. These give tasks and questions for the student to complete in their own time, to deepen and consolidate their learning. The questions may be straightforward with one-sentence answers, or they may be in the form of multiple-choice questions, completing missing words in a sentence, word searches and crosswords. They also include labelling diagrams and completing tables. Usually the answers to the questions are at the back of the book, so answers can be self-marked.

Many students have access to these books, or are even given them as part of their set reading books. However, even though they are invaluable as a resource, many students avoid using them. After all, a student needs to know what information they don't know. This begs the question, how can these student workbooks be included into the lesson plan?

Completing a workbook task at the end of each anatomy and physiology class could be a good way for the students to assess how much they have learned in that session. Alternatively, it could be included at the start of the subsequent session, to see how much knowledge has been retained. Chop and change the type of questions that are set – use a word search one day, a completion task the next, and so on. Encourage the student to follow up any wrong answers in their own time and learning space. Eventually, the student will be self-motivated to follow the workbook themselves.

A word of caution regarding word searches and crosswords ... Only use them to familiarise yourself with the terminology. Don't use them as your only source of revision or consolidation. There is a temptation to just simply find the word in the grid and mindlessly put a line through it, rather than actually think about what the word means. Follow up the word searches by using the list of words as a task to look up the definitions, or to test yourself on the meaning of the words.

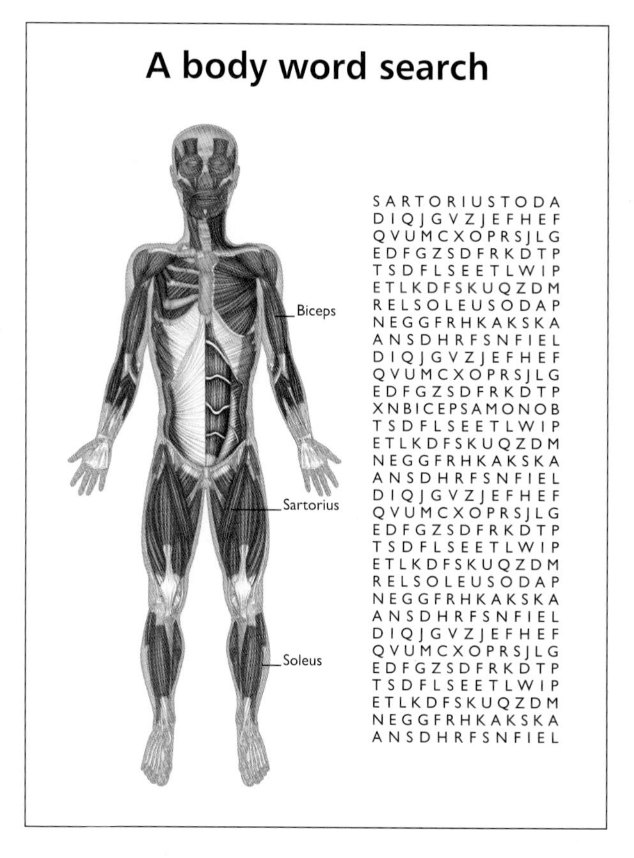

Figure 3.17: Example of an anatomical word search.

Kinaesthetic (with Visual) Learning

Kinaesthetic students are those who are the doers of the class. They like to get involved and take part in active learning, and will often be seen palpating/feeling their own bodies in order to work out where certain structures are located. All learners can take a leaf out of their book – after all, the only cheat you can take into an examination is your own body, so it makes sense to get familiar with it.

Using Our Own and Each Other's Bodies to Locate Structures

Locating and palpating individual anatomical structures on our own bodies will help reinforce the reading and lectures about a topic. Some structures cannot be palpated as they are too small or too deeply located in the body, but touching the body at the correct location whilst speaking the name of the structure will anchor in the memory its name and location, particularly for the kinaesthetic learner.

Finding a willing volunteer or 'study buddy' who gives consent is essential for the kinaesthetic learner. Practice finding the structures on someone else – they will be visually orientated in the same way as a diagram, so the usual issue of where is right and where is left is easier to overcome.

Think about the movements of joints and actions of muscles. Move the limb or structure and speak aloud the name of the joint or muscle and its actions whilst taking the body through the range of movement. Imagine you are taking an examination and cannot remember the exact origins, insertions and actions of a particular muscle as detailed in your course textbook? Think of the overall location of the muscle, and then mentally and physically bring its origins and insertions closer together; you will have moved the limb into its action. By then describing this action using anatomical language, you will be able to describe its action perfectly.

The sense of touch may be enhanced by closing one's eyes whilst palpating the structures. Removing the interference of other senses improves the sense of touch.

Get used to doing this throughout the course rather than leaving it until the final revision – the more the actions are animated, the easier it is to learn.

Anatomical Models

For teachers, the use of anatomical models is common practice. Pointing to a structure or allowing students to touch, hold and locate body parts is an accepted part of the course as it gives spatial awareness to the two-dimensional images in textbooks. Passing small models around the class is essential for kinaesthetic learners to feel and imagine the structures.

Visualising and feeling the three-dimensional image are important in understanding the locations and shapes of the anatomical structures. What is in front of or behind a muscle? How does the liver fit in that space above the colon and below the diaphragm? All can be answered and completely understood by using life-sized or two-thirds sized anatomical models. But how many students buy them for themselves?

If you have a birthday or Christmas coming up, ask for a present like no other. I have known a student receive a beautiful model of a spine, ribs and pelvis for a wedding present – I am none too sure what her husband thought of the gift, but she passed her exams with flying colours!

Whole life-sized skeleton models are great for looking at the big picture, but require space to stand the skeleton. Some people use their skeleton as a coat rack, or somewhere to hang their scarves, and the skeleton becomes a named member of the family. Meet Boris Bones ...

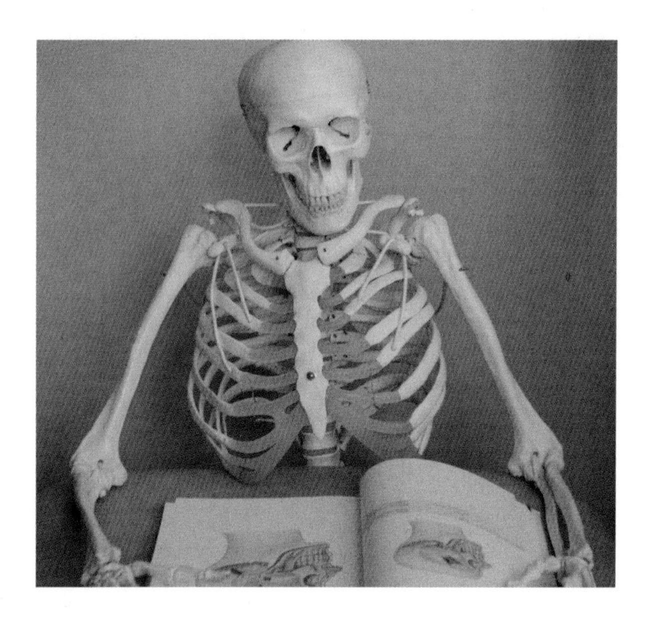

Figure 3.18: Meet Boris Bones, one of the staff of Learn Anatomy UK

Pharmaceutical companies often give out branded models of parts of the body; for example, the foot and ankle, knee, cranium or hand. These can be purchased quite reasonably in online auctions. They are often two-thirds sized and are perfect for labelling anatomical locations with a permanent marker or sticking on small labelled stickers. They can be easily packed away into a storage box, but equally they are small enough to leave out as talking points; talking about the models with friends and family will reinforce your learning.

Building Models

Creative students will enjoy building their own models from household objects. Teachers can guide the learners in explanation of the shape and size of anatomical structures, and structures can be built from a variety of equipment and resources, such as tubing, cotton reels, marshmallows, elastic bands, masking tape, electrical tape, straws, sponges, cling film and modelling clay.

Further examples of these can be found in Chapter 9 – Resource Anatomy.

Dissection

For many kinaesthetic learners, looking at a structure as a whole, then gradually revealing its layers and detail, will help them to remember the structure and location of anatomical parts. If learning within a medical school or university, this may be offered as part of the anatomy course, but most people who learn anatomy do not have access to cadaver dissection or autopsy.

Combining good dissection with thoughtful questioning will reap great rewards. Ask questions such as: describe how the stomach and the gall bladder are structurally similar or structurally different.

Prosection

Prosection is the dissection of a cadaver, either human or animal, by an experienced anatomist to demonstrate anatomical structure to students, either as a whole or part of a limb or structure. Some students really do find that seeing and feeling the anatomy in the flesh helps them to fully understand and appreciate the intricacy of anatomy (Theoret et al., 2007), whereas others may be squeamish. All human cadavers are from people who had given permission for such dissection and prosection.

Videos of prosection are available on the internet, but only offer visual information rather than the full kinaesthetic experience.

Animal Organs from the Butcher's Shop

If prosection or dissection is not an available resource, alternatives can be sought by befriending a good butcher. Looking at the structure and formation of particular muscles destined for the food chain may be more acceptable to many students, but obviously not all. Other structures such as the kidney, liver, stomach (tripe), tongue, long bones and eyes can all be purchased to dissect or examine at home. More about this in Chapter 4 – See Anatomy.

Imagination – Using VARK Together

Anatomy is to physiology as geography is to history; it describes the theatre of events.

Jean Fernel

Good teaching is one-fourth preparation and three-fourths theatre.

Gail Goldwin

What we learn with pleasure we never forget.

Alfred Mercier

By combining all our senses and using action and movement, we can use imagination to re-enact the forms and functions of anatomical structures. For ideas on how we can use imagination as a tool in learning anatomy and physiology, read Chapter 9 – Resource Anatomy and the Appendix – The VARK Questionnaire.

Making it Real

Though human ingenuity may make various inventions which, by the help of various machines answering the same end, it will never devise any inventions more beautiful, nor more simple, nor more to the purpose than Nature does; because in her inventions nothing is wanting, and nothing is superfluous, and she needs no counterpoise when she makes limbs proper for motion in the bodies of animals. But she puts into them the soul of the body, which forms them that is the soul of the mother which first constructs in the womb the form of the man and in due time awakens the soul that is to inhabit it.

Leonardo Da Vinci

The majority of people learning anatomy do so in order to go on to study medicine, pathology, physiotherapy, nursing, biomedical science, veterinary science, therapies of some kind (beauty or holistic) or similar, where they will be expected to apply the knowledge learned in the anatomy and physiology class. It therefore follows that making the anatomical knowledge more appropriate to the career choice will make it easier to learn, understand and retain, as well as being far more useful.

Diseases and Symptoms

Anatomy and physiology detail the structure and function of the healthy living creature, but, in reality, the information that is more relevant is what happens when it goes wrong. When the body's structure and function is compromised, and the body is unable to use its homeostasis to bring it back to normality, disease results. The signs and symptoms of disease are usually more memorable and more interesting to most people than bald facts.

Work Backwards from Diseases and Symptoms, thus Highlighting Structure and Function of Anatomical Features

Consider how a body would function if a body part were missing – what symptoms would arise? What would the body be unable to do? Then look back at the functions of that structure and link together the functions with the dysfunctions that arise in the absence of a structure. Imagine what a leg would be like without bones.

Figure 3.19: How would a body function without a body part?

Consider a specific disease – for instance, renal failure – and list the symptoms of the disease. Then look back at the functions of, in this case, the kidney, and look at how the symptoms of renal failure are due to dysfunction (poor function) of the kidney.

Case Studies (for Pathology, Medicine and Holistic Therapies)

Looking at specific diseases and conditions as case studies is a useful way to make the anatomy real and therefore more memorable. Bringing in patients with known conditions is extremely memorable, particularly if the patient is able to answer questions from the students.

If you are a student of anatomy, and you actually know someone with a medical condition that you are studying, do your best to imagine and remember their symptoms and compare them to the known functions of the body systems affected. Make it real!

Practical Experiments

Classroom practical experiments using live subjects (other students or volunteers) will be a memorable way of enlivening anatomy and physiology learning. Examples of this could be:

- Effect of exercise on heart rate – cardiovascular system
- Respiratory volume at rest and during exercise – respiratory system
- Colour blindness tests – visual system
- Simple reflex testing – nervous system.

Taking Responsibility for Learning

Spoon feeding in the long run teaches us nothing but the shape of the spoon

E. M. Forster

The aim of every teacher should be to empower the students so that they are taking responsibility for their own learning. In this way, the learner becomes his or her own teacher.

'See One, Do One, Teach One, Practise'

In order to fully understand a subject, it can be useful to adopt the 'see one, do one, teach one, practise' technique. Seeing allows the student to be taught a piece of information, doing allows the student to activate the learning and teaching allows the student to pass on the information to a peer or a family member. It is often in the delivery of teaching a peer or family member that the realisation dawns that the topic is not fully understood. Repetition and practice further reinforce the learning, thus allowing the student to 'own' the information for him or herself.

Peer marking and peer mentoring also use a similar tack.

> *In learning you will teach, and in teaching you will learn.*
>
> Phil Collins

Peer Learning in Small Groups

'Study buddies' or peer groups are an effective way of sharing anatomy knowledge. As we have seen, everyone learns in a variety of ways, so study buddies should either learn in the same way, or complement each other's learning. Going through recent class material as a group will help to share the load, and often one person may have mastered a topic that others have not properly understood, so the 'see one, do one, teach one, practise' technique can be used.

Always have an agenda for a small group session. All too easily, the group can descend into self-doubt and a whinge session about how much they don't understand. Use anatomy games and other methods of learning mentioned in this chapter to structure the small group. Invest in large flip-chart paper to draw big diagrams that can be labelled as a group.

Some classes are rather geographically challenged, with students living too far away to actually meet up. Modern technology means that, via email or internet, conversations can take place easily. Use social media to set up an online group or forum. Make sure that privacy settings are used to keep the content and membership completely private. Use this to share questions and topics regularly. Help each other. Perhaps take it in turns to post a daily factoid about anatomy and physiology. Post potential examination questions for the other students to attempt to answer.

Final Thoughts

Learning any subject can be a challenge, particularly if an inappropriate learning style is used. Matching the learning style to the learner and creating memorable ways of approaching the teaching and learning of anatomy and physiology will help the subject to make sense, and be integrated and applied into career paths. Just turning up at lectures and reading a textbook is an ineffective way of learning anatomy and physiology – you have been warned. Be creative and experimental in learning styles, and be prepared to step outside comfort zones in order to enrich both teaching and learning.

References

American Red Cross, *Fundamentals of Instructor Training*. Available online at: http://youth.net/ecc/FIT.pdf. (Accessed 16 January 2016.)

Bolte Taylor, J. (2008). *My Stroke of Insight: A Brain Scientist's Personal Journey*. Hodder & Stoughton.

Stice, J. (1987). Using Kolb's learning cycle to improve student learning. *Engineering Education* **77(5)**: 291–296.

Theoret, C.L., Carmel, E.N. and Bernier, S. (2007). Why dissection videos should not replace cadaver prosections in the gross veterinary anatomy curriculum: results from a comparative study. *Journal of Veterinary Medical Education* **34(2)**: 151–156.

VARK Learn Ltd (2015). *Research and Statistics*. Available online at: http://vark-learn.com/introduction-to-vark/research-statistics/. (Accessed 14 January 2015.)

VARK Learn Ltd (2015). *VARK Profiles Jan–March 2015*. Available online at: http://vark-learn.com/wp-content/uploads/2014/08/varkPreferencesGraph2.gif. (Accessed 14 January 2015.)

4 See Anatomy

This chapter considers the visual nature of learning anatomy and physiology. We acknowledge that learners use multiple methods to learn and that not all students are visual learners. In deciding on learning style it may be useful to consider the VARK assessment. This is a test to assess dominant learning style and is useful in providing an analysis of different learning styles ranging from visual to auditory to reading/writing and, finally, kinaesthetic awareness. This chapter's content has been aligned with the content of Chapter 6 – Learn Anatomy, and the two chapters should be read in conjunction. The VARK scale is further explained in Chapter 3 – Do Anatomy, and the Appendix – The VARK Questionnaire. Throughout this chapter we consider how best to make connections and create building blocks for seeing anatomy in action.

> *The inner image we remember from a process is not the result of our perception of the same thing or process.*
>
> Jean Piaget, 1966, *The Psychology of the Child*

> *Vision is the art of seeing what is invisible to others.*
>
> *Jonathan Swift*

Anatomy is often described as a visual study, one which requires creative images and drawings to bring to life the content we read. Imagine a textbook of anatomy and physiology that described structure and processes devoid of any form of visual illustration. Some may find reading the content difficult, others would argue that visual images could distract the reader or learner when having to recall facts and processes. Whichever method you choose to adopt, when studying anatomy we cannot escape the fact that the study and dissection of the body comes to life through seeing how the parts connect and relate to each other.

I See – But Do We All See the Same Thing?

Drawing on my own experience of studying anatomy, I realised early on in my career that, for me, anatomy was one giant puzzle wherein I needed to know how structures form and perform, especially when the body is in a diseased state. I vividly recall the day I was invited to watch a family friend and recognised orthopaedic surgeon perform a total hip replacement on an elderly woman who had osteoarthritis. Standing in the operating theatre and peering into the incision the surgeon had made to expose the tissue structures made me feel queasy but extremely fortunate for having witnessed the human body inside out. I watched carefully as, layer by layer, different tissue was dissected. Seeing the body through cut sections made me realise that textbook drawings are neat and colourful but do not really depict the body from multiple views and perspectives.

Drawings are useful, yet seeing the human body from a surgical perspective, with its interlaced networks of muscle, fascia, bone, blood vessels and nerves, created a whole new dimension of complexity. The surgeon proceeded to quiz me on the anatomy of hip, pointing out minute structures. This provided a wonderful canvas for me to orchestrate my new understanding of the human body and create a 'learnscape' for myself and my students.

I was so fascinated by seeing the human body in differing views and angles that I took a group of students to see Gunther von Hagens' Body Worlds exhibition. As we entered the exhibition we were confronted with a giant dissection of a horse, separated and compartmentalised. This precision dissection permeated and punctuated the exhibition. My students were able to visualise how complex structures form, combine to support the body and network with other structures to sustain life. Within the exhibition there is a wonderful dissection that includes a dissected man holding up his own skin. My students stopped and stared at this particular specimen and were fixated on the levels and layers of soft tissue and their connections with joints and mobility. When we returned to the class, I set my students a task of examining the human body through the work of artists. They were asked to consider how artists viewed the human body and its workings and how their understanding was depicted in their artwork. Many of my students were perplexed and confused with this assigned task. A week later I received their work and was pleasantly surprised to see how they had reflected upon their experience at Body Worlds and presented an account of the human body from both a scientific and an artistic perspective. This task enabled my students to appreciate the visual nature of learning anatomy and how seeing structure in action can influence perspectives and develop deeper understanding.

When I was a student trying to ingest anatomical terms and concepts, I had to rely on my own understanding of art and drawing. This was before the age of the internet, where finding images and videos of the human body is simply reliant on the click of a mouse. The internet is flooded with useful, and also not so useful, animations and graphics to illustrate structure in relation to function (see Chapter 9 – Resource Anatomy). Just like reading and rereading content, anatomy as a visual study relies on watching and observing. In Chapter 2 – Speak Anatomy, we used the example of reading from a book and then re-examining content to extract meaning. The same thinking and analogy can be applied to watching a film. The first time we watch, we look for certain features; however, the second and third viewings reveal elements we failed to see the first time. This encourages us to connect and reconnect with the content we watch and study. If asked whether you could recite your favourite line from a film, you probably need to replay the scene in your head as you identify the line with the character and situation in which it was said. Anatomy is similar in design, wherein we learn to focus our learning on relating specific images to functional understanding.

Today my fascination for anatomy transcends into my love for cooking and preparing food. I often show my young children different anatomical structures as I prepare their meals. They too are interested to see how joints articulate and how blood flows. Instilling this love for the study of living things at a young age is exciting. I hope they grow to appreciate how best they can maintain a healthy and balanced lifestyle to minimise disease and ill health.

Do we all look for the same things when we learn anatomy? So what is it that we see or fail to see? Below is an image from Gestalt psychology where foreground and background create different imagines. Look at the image in Figure 4.1 and decide at first glance what it is you see. Then re-examine the image to see if you see a different or more detailed image.

We either see an elderly couple, or Mexican musicians. What we see is dependent on how we look and choose to see.

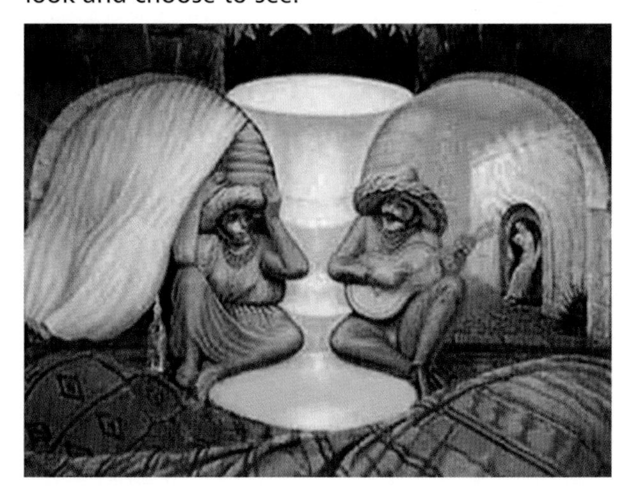

Figure 4.1: Image of elderly couple, or Mexican musicians.

Taking this further, how many legs does the elephant in Figure 4.2 have?

Figure 4.2: An elephant – count the legs!

Our brains learn to fill in the gaps and complete the picture. This ultimately could lead to confusion and the creation of an obscure or abstract image.

How do we make anatomical knowledge stick so that it can be remembered long after we have closed the chapter or book?

How Do We See?

Eying out the Brain

Seeing is a complex conjugate between the eye or ocular tissue and the processing mechanism within the occipital lobe of the brain. The eye functions to control light. Images we see are processed initially in the eye, but then travel via the optic nerve to the occipital lobe, where an inverted image is processed and interpreted. The interpretation of what we see is predominantly the role of the brain. The brain creates an awareness of what we see, whereas the eye creates the image for this awareness. We interpret colours, shape, textures and contrasts to ensure that what we see is clearly defined and meaningful.

Humans have learned to adjust images to depict processes, events and occurrences. It is important to consider not only how we see, but, more importantly, what we see and learn. Have you ever been to an art exhibition? Typically, this experience enables people to stand in a room and stare at a picture on the wall. I often consider the behaviours of individuals when they look at artwork and then engage in discussion about the meaning and interpretation of the art itself. There is a cognitive cascade of ideas about how the artist felt when painting the picture, and how the picture symbolises complex emotional states. The simple process of looking, watching and observing interacts with higher-level brain functions of emotional analysis. Imagine, if you will, just for a few seconds, that you are invited to an exhibition on the human skeleton. In this particular exhibition there are over 50 skeletons positioned and captured in different ways. How would you react to seeing this exhibition? What learning or discussion would you draw from what you see? How would this opportunity to see these skeletons influence your ability and understanding of learning skeletal anatomy?

Using Vision for Learning

There are two main images we learn to recognise and use. The first is a reproduced image. For this image, we need to make judgements and comparisons to previous sight of the image. We learn to recall the perceived image and map it to a content of learning. The second image is slightly more complex and involves visual change and anticipated outcome independent of direct perception. This is called the anticipated image and is used to predict what we see in motion.

In the study of anatomy we learn to use both image types for different functions. In the reproduced image we learn to compare what we know with what we see or read. There is a constant juggle between seeing, understanding and drawing meaning. In the second image we learn to apply analytical skills to help bridge gaps between static and dynamic motion. We learn about the skeleton, for example, in a single plane, but then need to develop an understanding and appreciation for human movement by observing skeletal changes during the walking process. This involves observation of the gait cycle and integration of biomechanical knowledge.

Imagery is a powerful psychological skill, used to aid our understanding of our learning, reflective practices, and awareness of feelings and thoughts. There are multiple forms of imagery, such as:

- *Visual imagery*, pertaining to graphics, visual scenes, pictures or the sense of sight
- *Auditory imagery*, pertaining to sounds, noises, music or the sense of hearing
- *Olfactory imagery*, pertaining to odours, scents or the sense of smell
- *Gustatory imagery*, pertaining to flavours or the sense of taste
- *Tactile imagery*, pertaining to physical textures or the sense of touch.

The way we use imagery is important for how we choose to manage our learning. Poetry is a tapestry of imagery, created through rich blends of descriptive metaphors. Anatomy, similarly, builds imagery through aligning function and dysfunction with structure and morphology.

Many students use imagery to revise what they have learned. This could involve lying in a quiet place and running a silent movie through the corridors of their minds. The student is the engineer, director and choreographer of the images he or she views, learns and studies. Through imagery one can dissect complex structures into smaller manageable components so that the learning becomes sequential and meaningful. Breaking a film down into component parts allows the individual to focus more clearly and deeply on specific issues and events. If we simply focus on the entire process, we often miss the details in each segment. If you were to image how we digest food, you could start at the mouth by describing structures, progress onto the oesophagus, stomach, intestines, rectum and anus. It may be more important to spend time in each component to fully appreciate how the structure relates to function, and how the parts form the whole – i.e. from dissection to resurrection. Further illustrative examples can be found in Chapter 7, on preparation for assessment.

An imagery script is a useful learning technique that enables students to develop learning through words, images and association of ideas. The script can be developed over time, with key learning concepts and phrases embedded into the context.

Imagery Script – the How To Do

In preparing an imagery script for learning it is important to embed and consider a clear image of the content to be learned.

The script is then developed by understanding:

1. The key terms for the concept or structure (language of anatomy)
2. Function of the structure
3. Applications
4. Integration
5. Purpose
6. Motivational images and learning cues to inspire learning
7. Further images
8. Connection between images and concepts.

The script, if produced effectively, has the potential to create a learning space that allows the student/ learner to develop knowledge through illustrations and associations. This builds the foundation for future learning and inquiry.

Creating Connections

Our brains have a wonderful internal network for creating and maintaining connections. It is useful to think of the nerves as a communication system for relaying information to and from the brain. When we view images we create a collage of thoughts and associations that cascade to produce a meaningful account of what it is we see, or, often, fail to see. Our response to an image or visual cue originates from a stimulus. The stimulus activates cells in clearly defined areas in the brainstem. This leads to changes in the body such as narrowed attentional focus or enhanced fixation on the image. Finally, information about the image is transmitted to the brain where evaluation of the image itself occurs, i.e. what is it that I see? Or, how do I connect what I see to what I have learned? By using different images and colours we learn to recognise features within images. Should such an image appear in an examination paper, we can use our prior understanding of the image to find answers to questions about the image or diagram.

Figure 4.3: Converging thoughts producing a meaningful image, much like interconnecting threads producing a rope.

The next section provides information about how to use visual learning to study or revise anatomy and physiology content.

Methods for Enhancing Visual Learning

Mind the Gap – Understanding and Developing Mind Maps

A *mind map* is a diagram or scheme used to visually organise information. A mind map is often created around a single concept, drawn as an image in the centre of a blank landscape page, to which representations of associated ideas such as images, words and parts of words are added. Major ideas are connected directly to the central concept, and other ideas branch out from these ideas. Mind maps have typically been used to plot and direct concepts and applications in the study of anatomy and physiology. They enable the student and teacher to consider associations and connections between ideas and how these ideas connect with other concepts and processes. The mind map gives the user a journey chart to navigate the learning and signpost important areas for learning and consolidation of learning. Why not try making a mind map of a topic in one subject area, using different colours for each part of the topic, and putting down the key words that will help you remember? It really is worth trying this – if your brain works this way, mind mapping is a perfect way of remembering huge amounts of information. It may not work for you but you do need to practise a little before you can make that decision.

Why Mind Map for Anatomy?
Mind mapping can assist with:

- Revision notes
- Learning anatomy as a language
- Goal planning
- Looking at connections between terms, concepts and ideas.

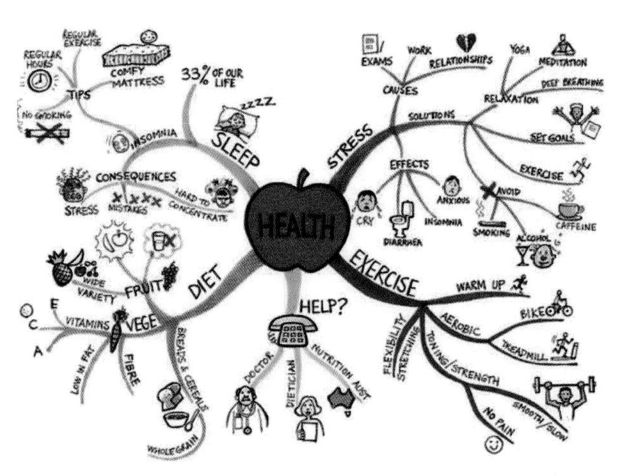

Figure 4.4: A mind map.

How To Develop a Mind Map

- Consider the topic you wish to mind map, e.g. the flow of blood through the heart. What will the focus of your thinking be? Use few words to generate the topic.
- Generate a draft using different colours and paper.
- Use the centre of the page first.
- Size really does matter here – mind maps should be no larger than an A4 sheet of paper. You could also consider the size of the images and text you use to emphasise ideas, processes and concepts.
- Association of ideas – record what comes into your thought and visual field, draw branches from the main topic.
- Idea fusion – use words such as what, why, where, when and how to help shape the mind map.
- Limit choice of words – use single words or phrases to capture ideas.
- Branch out – extend thoughts and ideas from the main topic and associated branches.
- As new ideas flood in, continue to draw branches and make associations between concepts.
- Repeat the branches until all the ideas are on the map; this provides a recipe for your visual learning of a set concept, process or organisational activity within anatomy.
- Reflect and connect – refine connections; are there any larger patterns or associations?

Through the mind map, one begins to create a storyboard for anatomy; just as the director of a film needs to carefully consider the positions and interplay of scenes within the film, so too do students and teachers need to consider the positions and interplay of anatomical structures. A mind map becomes the vehicle through which we learn to conceptualise what it is that needs to be taught, studied and learned.

The Impact of Colour

Colour can enhance and relay mood states. It enables students and teachers to emphasise key facts and features within the study content. Through colour we can better adjust our thinking and planning to areas for focus and development. Anatomy is a visual study, one which demands the use of colour for effect and emphasis. This is evident in anatomical texts, charts and learning resources. Anatomy devoid of colour is akin to a landscape devoid of life. Through colour we learn to differentiate structure, label and identify features, and create connections between structures and systems. As infants, we are drawn to or repulsed by certain colours. We learn to communicate through colour before our language skills are developed. In fact, colour connects the right and left hemispheres of the brain, allowing both gestalt and analytical learners to interact and master specific tasks. The one question most often asked and the least answered is, 'What is the colour of learning?' Is there such a thing as a 'learning colour'?

The table provides some insight into the power of colour, together with the emotional effect. The table has been developed from the work of Wright (2015) and KD Web Ltd (2011), web designers. According to Wright, there are four psychological primary colours, namely red, blue, yellow and green. These colours relate to the body, the mind, the emotions and the homeostasis between the three.

Colour	Description	Emotions
Red	Red is strong, and very basic. Pure red is the simplest colour, with no subtlety. It is stimulating and lively, very friendly. At the same time, it can be perceived as demanding and aggressive.	*Positive*: physical courage, strength, warmth, energy, basic survival, 'fight or flight', stimulation, masculinity, excitement *Negative*: defiance, aggression, visual impact, strain
Blue	Blue is serene and mentally calming. It is the colour of clear communication. Time and again in research, blue is the world's favourite colour. However, it can be perceived as cold, unemotional and unfriendly.	*Positive*: intelligence, communication, trust, efficiency, serenity, duty, logic, coolness, reflection, calm *Negative*: coldness, aloofness, lack of emotion, unfriendliness
Yellow	The right yellow will lift our spirits and our self-esteem; it is the colour of confidence and optimism. Too much of it, or the wrong tone in relation to the other tones in a colour scheme, can cause self-esteem to plummet, giving rise to fear and anxiety.	*Positive*: optimism, confidence, self-esteem, extraversion, emotional strength, friendliness, creativity *Negative*: irrationality, fear, emotional fragility, depression, anxiety, suicide
Green	Green strikes the eye in such a way as to require no adjustment whatever and is, therefore, restful. Being in the centre of the spectrum, it is the colour of balance – a more important concept than many people realise.	*Positive*: harmony, balance, refreshment, universal love, rest, restoration, reassurance, environmental awareness, equilibrium, peace *Negative*: boredom, stagnation, blandness, enervation
Orange	Since it is a combination of red and yellow, orange is stimulating and reaction to it is a combination of the physical and the emotional. It focuses our minds on issues of physical comfort.	*Positive*: physical comfort, food, warmth, security, sensuality, passion, abundance, fun *Negative*: deprivation, frustration, frivolity, immaturity
Grey	Pure grey is the only colour that has no direct psychological properties. Heavy use of grey usually indicates a lack of confidence and fear of exposure.	*Positive*: psychological neutrality *Negative*: lack of confidence, dampness, depression, hibernation, lack of energy

Black	Black is all colours, totally absorbed. The psychological implications of that are considerable. It creates protective barriers, as it absorbs all the energy coming towards you, and it enshrouds the personality. Positively, it communicates absolute clarity, with no fine nuances. It communicates sophistication and uncompromising excellence and it works particularly well with white. Black creates a perception of weight and seriousness.	*Positive*: sophistication, glamour, security, emotional safety, efficiency, substance *Negative*: oppression, coldness, menace, heaviness
Pink	Being a tint of red, pink also affects us physically, but it soothes, rather than stimulates. Pink is a powerful colour psychologically. It represents the feminine principle, and survival of the species; it is nurturing and physically soothing.	*Positive*: physical tranquillity, nurture, warmth, femininity, love, sexuality, survival of the species *Negative*: inhibition, emotional claustrophobia, emasculation, physical weakness
White	Just as black is total absorption, so white is total reflection. In effect, it reflects the full force of the spectrum into our eyes. Thus it also creates barriers, but differently from black, and it is often a strain to look at. It communicates, 'Touch me not!' White is clean, hygienic and sterile. The concept of sterility can also be negative. Visually, white gives a heightened perception of space.	*Positive*: hygiene, sterility, clarity, purity, cleanness, simplicity, sophistication, efficiency *Negative*: sterility, coldness, barriers, unfriendliness, elitism

Colour not only invites creativity, but further establishes boundaries between and within ideas and concepts. The next time you use colour, think carefully what that particular colour means, what it says about emotional states, and how best it can be used to enhance and promote learning. The more colours we use with difficult concepts or troublesome learning, the easier it may become to devolve learning into manageable clusters.

Earlier in this chapter we introduced the concepts of imagery and visualisation. Using the principles of imagery, can you think of a vivid example where colour made a difference in your life? It may have been a newly painted or decorated room; a new gadget, car or watch, where the colour relayed a stronger message than words could have done. We often become oblivious to the power of colour. How many times have you used highlighter pens to mark off passages of text during a revision session, often using multiple colours to emphasise and re-emphasise key learning points? Subconsciously our brains are working to interpret these visual explosions so that the content is stored and retrieved when required. It is through colour that we begin to appreciate the complexity of the arrangements within our remarkable human body.

Think about when you have viewed images of microscopic cells and structures. Think about the colours of the nucleus, the cytoplasm and the collagen. Those colours are not there naturally, but are dyes and stains chosen by biomedical scientists or histologists to highlight certain structures to the eye when viewed down the microscope. Without these stains and colours, the structures would just be pretty transparent. The routine stain of choice for microscopic viewing of human cell structure or 'morphology' is haematoxylin and eosin (H&E). This process uses a haematoxylin dye to stain the nuclei of the cell a bright blue, and all the other structures are stained pink/red by the eosin. It is the main stain used in hospital histology laboratories for tissue-based diagnosis, and the detection and classification of cancer. Microscopic images of blood cells have usually been stained with a two-part stain of May-Grünwald and Giemsa at a defined pH (usually pH 6.8), giving the effect of purple nuclei and a variety of shades in the red to blue spectrum. Other poetically named specialised stains can be used to visualise specific microscopic structures: Masson's trichrome for collagen fibres; Perls' Prussian blue for ferric iron in tissues such as bone marrow and spleen; Gomori trichrome for muscle fibres and collagen; and periodic acid–Schiff for structures containing a high proportion of carbohydrates such as glycogen and proteoglycans, which are typically found in connective tissue and basement membranes. Next time you look at a photograph of a microscopic image of cells, just think about the preparation techniques that have been used to allow you to see them.

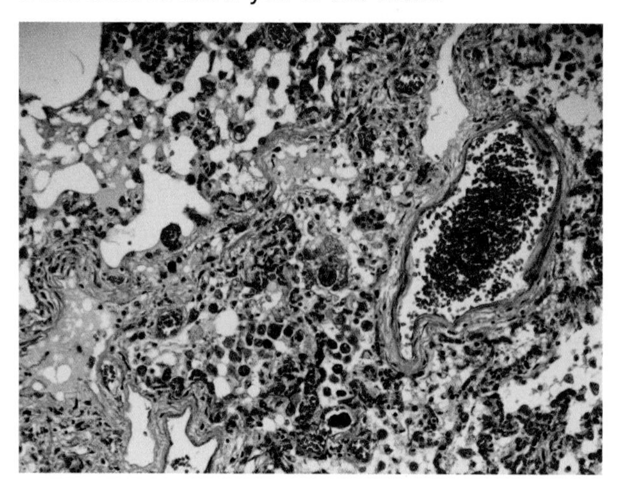

Figure 4.5: Stained cells – collagen in the blood vessel appears blue when stained with Masson Trichrome. (Photo kindly supplied by West Suffolk Hospital NHS Trust Histopathology Department.)

By using colour as an expression of understanding, anatomy is transformed from scientific study to an art form. Imagine a tattoo artist painstakingly etching the intricacies of the human form on an individual's skin. The artist engages the process of visualising where structures lie in relation to other structures and how best to depict three-dimensional shapes on a two-dimensional human canvas. The ability to draw and use colour by intuitively knowing how colours complement each other, to produce shape and form, is a wonderful dimension of learning and analysing the study of the body.

In preparing this chapter, I looked through the numerous anatomical textbooks I have read and used as a student many years ago. The one book which partly inspired the development of this chapter is *The Anatomy Coloring Book* (Kapit, 2002). The book is modelled on a child's fascination and intent in colouring in pictures and shapes. Through mastering how colours support and strengthen an image, the anatomy colouring in book communicates a stronger message of *learning through doing*. The book provides a platform for students to use colour and differentiate structures. The final product represents the anatomical component in an art form which shapes learning and application. For example, by using similar colours to group structures of similar function, a toolkit is created for identifying the structure within a group and then considering how the group works to produce action and outcome. It is important to remember the purpose of colouring in. Students/learners tend to focus on staying in the lines and producing beautiful artwork. The aim of colour is to emphasise learning and understanding, not simply to produce colourful and artistic pictures.

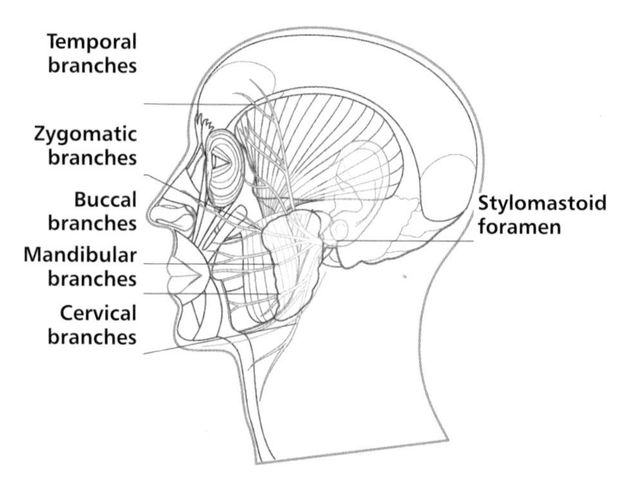

Figure 4.6: An anatomical structure for colouring in.

The simple and effective application of colouring in anatomical structure has allowed for creative developments such as anatomical colouring shirts, where students learn to ensure the accuracy of their art, which is then worn as a reminder of their learning and revision. The anatomical clothing idea provides practical evidence and support for using images to remind students of how the body is organised and how structures work in unison to orchestrate movement and function. It is useful to either predesign the structures that students could colour, or provide the students with a blank canvas for them to design their own anatomical shirts. There are numerous variations on this concept, for example body painting (refer to Chapter 9 – Resource Anatomy for further ideas).

Body Painting – Not Simply a Case of Draw Backs

Body painting has numerous learning benefits, such as in the teaching and learning of the musculoskeletal system. This hands-on method affords the student an opportunity to practise multiple skills in preparation for, and application of, artistic interpretation of structural components.

Figure 4.7: Body painting (Kathy Dooley, Anna Folckomer and artist Danny Quirk).

The Process

Students receive a cue card with an image depicting a muscle. They then need to apply their palpation skills to locate bony markings and anatomical points of reference. Once these have been considered, the student then draws the outline of the structure by carefully noting its relationship to other regional structures. For example, if a student was to draw the gastrocnemius muscle, he or she would need to consider the number of heads, the number of joints the muscle spans, the muscle action in relation to the pull of its fibres, whether it is superficial or deep, or whether it bars vision of other muscles superficially. Students could use different colours to differentiate superficial from deeper muscles or extensors from flexors. There are numerous permutations that could be used during body painting, it is only by doing and experimenting that we truly learn to recognise and appreciate the endless options and learning opportunities.

I recall once having an arts graduate attend one of my practical anatomy classes. She was fascinated with how anatomy could be expressed artistically. I remember watching her in the body painting session, where she spent two hours painting three lower-limb muscles. When I eventually saw the final product I quickly realised why she had taken so long. She had painted each muscle fibre, and blended the muscular and fascial systems by locating and outlining bony prominences as points of muscle attachment. This experience of observing an artist at work allowed me to step back and reflect upon how visualisation becomes a lived experience – one wherein the physical dimensions of the human body, often concealed under layers of tissue, become an external expression of a need to understand how things work. The other students photographed the arts graduate's artwork and used it as a key revision tool. Today, with the advent of smart technologies, one is able to photograph and upload the artwork to applications that embed the art into a learning technology platform, thereby creating a galaxy of further learning opportunities.

Wallpaper

One need not be an artist to study or learn anatomy; however, using canvases and drawings is useful in positioning content within a learning framework. One technique which is commonly used by students is to draw images and use mind maps on huge sheets of paper (see Figure 4.4). These sheets are then plastered onto walls as a constant reminder of content and application. One of my students drew an intricate tube map to depict similarities between the tube lines and the brachial plexus. The idea of the brachial plexus being synonymous with different train lines, enlarged on a canvas spanning an entire bedroom wall, brought to life the intricate connections between nerves and soft tissue, similar to the relationships between trains and stations. The student was able to use colour to dissect and chart the course of each individual nerve within the plexus. This further enabled her to better understand nerve innervations and muscle actions.

The wonderful advantage of using wallpaper as a learning technique is the nature of the wallpaper itself, which is ever changing and evolving. This gives students the opportunity to identify gaps in their learning and amend mistakes, but most importantly build upon a foundation of knowledge. Wallpaper can now be transposed onto a computer desktop, which further allows for recognition of learning whilst developing new knowledge and applications. The more visually exciting the wallpaper is, the more the student will want to review it on a regular basis. Having said this, the content of the wallpaper is highly subjective and dependent upon the student's unique and/or integrated learning style.

Anatomy is Transparent

Drawing and visual art are exciting techniques to bring to life anatomical and physiological content. A technique which has powerful effect is the use of transparencies to layer the human body. Students can use a number of transparency sheets to detail different anatomical layers and connections. This enables the students to further consider the relationships within and between connective tissue layers, body cavities and anatomical boundaries. It provides an observation lens to scrutinise internal and external processes. The advances in transparency art have recently led to the development of layered tattoos that encourage the user to peel away the upper layers to discover deeper anatomical structures and processes. This is a fascinating art form that engages the learner with the learning content and creates an experience of understanding the function of complex body organisations. This form of art invites the learner to see through the structures and visualise how body control is maintained and monitored, and how one part of the body has an impact on and affects the whole.

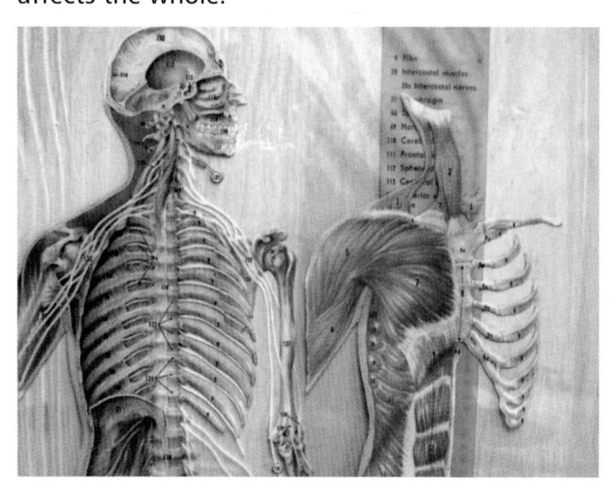

Figure 4.8: Transparencies.

The Comical Side of Anatomy as a Visual Study

I have often watched young children learning to read. They tend to build the story based upon the pictures that support the story content. For them, the pictures tell the story, not necessarily the words. The study of anatomy is not too dissimilar in that new learners tend to focus on diagrams and visual cues to interpret meaning as opposed to reading lengthy text that may confuse rather than assist their learning. With the advances of the internet, students can now watch video clips that explain, through animation, complex phenomena and concepts. The visual advantage of seeing as opposed to reading may be to instil a greater and deeper appreciation for the content of the learning. One technique which allows students to explore their imaginary and creative sides is the creation of cartoons and comic strips to support their learning. With this technique students can explain their learning and understanding of concepts through cartoons and animations by creating characters and scenes to depict a specific process of events. An example is the use of cartoon characters to illustrate the digestive process, aligning the mouth with a compressor unit, the stomach with a cement mixer and the colon with a plumber's pipeline. The cartoon emerges as an anatomical take on Bob the Builder, or, anatomically speaking, Bob the Bolus!

In developing a cartoon or comment sketch it is important to consider number of scenes, characters, facts, illustrations, and flow of events or storyline. The cartoon images need to be meaningful for the learner, the facts need to be accurate and the flow of events needs to be considered as a sequenced process. This process can then progress into short videos and animations.

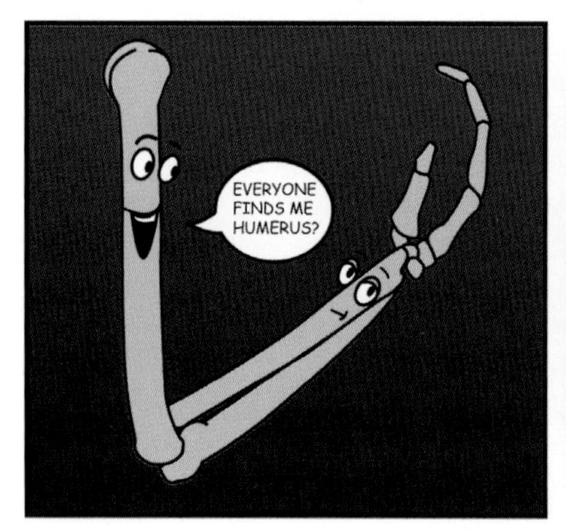

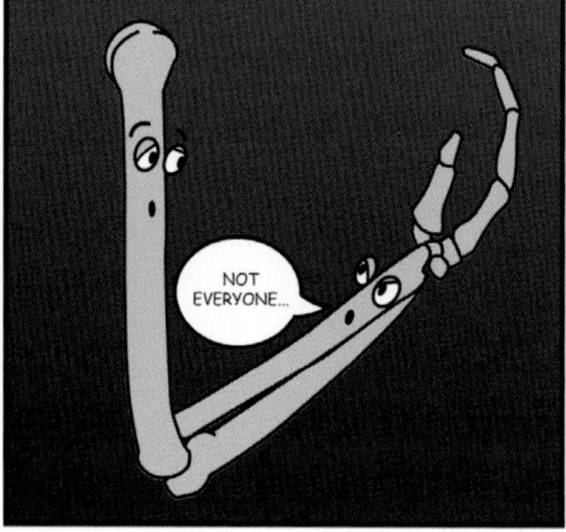

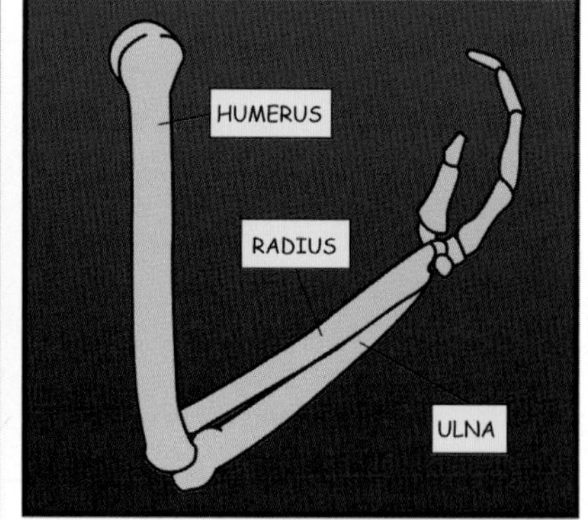

Figure 4.9: A cartoon strip with an anatomical theme.

Games and Puzzles

Throughout this text we have referred to the power of games and puzzles. A number of chapters detail and explain the value of games in learning and concept expansion. The numerous points in the text that have mentioned games and puzzles are positioned to accentuate and emphasise the ideology behind the games.

There is a lot to learn from children's games and puzzles. The wooden age has evolved into a digital one filled with online games, quizzes and problem-solving activities. Jigsaw puzzles have evolved into complex designs of three-dimensional images that need to be pieced together to form the whole. Anatomy lends itself to the use and design of puzzles and games. Students use cue cards with visual images to revise key learning concepts. The design of puzzles enables the student to consider how parts fit together to form a structure and how the structure functions to maintain the whole. Puzzles and games are only limited by the creativity and the imagination of the creator. A simple technique is to produce a diagram or illustration of an anatomical structure and then divide it into a number of smaller pieces. One could further create some visual cards to illustrate a process or ask the question of what happens next. Games such as Pictionary™ are useful in encouraging development of ideas and integration of knowledge.

Meat Your Needs ...

At the beginning of this chapter, I alluded to my passion for cooking and preparation of food. I would like to draw this chapter to a close by considering the use of food products to aid visual learning.

Of all the individuals I have learned to love as part of the anatomy learning community, my butcher is at the top of the list. My request list for joints, organs and prosections has surpassed any of his regular clients' requests for meat. I have learned to forge good working relations with my butcher and have discussed at length my teaching aid requirements. My butcher has over the years, provided wonderful cuts and cross sections of structures that I use to illustrate structure and function. It is important to consider ahead of class time the visual requirements for a lesson. Butchered off-cuts are often an inexpensive yet highly effective way to illustrate component parts. Students can use these aids to reinforce their learning or feel what supporting structure feels like based on the structural design. To retain the longevity of animal specimens such as bone, one could use varnish. Remember when one cooks or prepares food there is always a learning aid close at hand, for example chopping up carrots could be used to depict retinal arrangements, cauliflower could be used to show the organisation of the grey and white matter of the brain, whereas grapes are useful learning aids to depict the arrangement of the alveoli and infundibuli. The more I cook, the more I become engaged with developing food related anatomical learning aids.

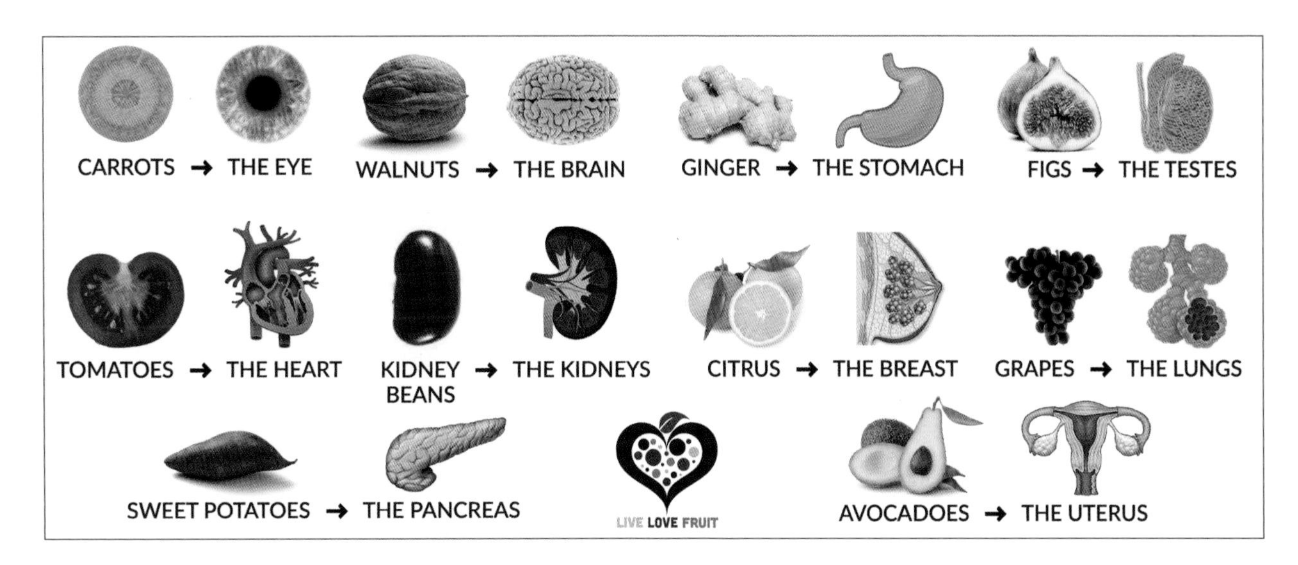

Figure 4.10: Anatomy in food. (Image courtesy of livelovefruit.com.)

Eat Your Heart Out

The visual and confectionary study of anatomy can be entertaining. A student of mine threw an anatomy-themed party with anatomically correct cupcakes of major organs and structures such as the heart, brain and lungs. Each cupcake was intricately iced with major vessels, chambers and coverings. As we partook in the food, we could not help but consider what it was we were eating, but, more importantly, how we learned whilst we ate. It is beyond the scope of this book to provide recipes for cupcakes, but Chapter 9 – Resource Anatomy contains interesting links to novel anatomical stores where one can purchase moulds to bake such cupcakes and confectionary delights.

Final Thoughts

This chapter has provided an overview and understanding of anatomy and physiology as a visual study. At times, it is important for teachers to help their students/learners see things such as structure, shape and organisational tiers. It is only through knowing how to see that we truly begin to see the components that become important in the learning and application of anatomy and physiology as a lived study. The study of anatomy and physiology creates opportunities for direct looking into, seeing and appreciating what it is we learn to see. This helps us to make sense of the body and appreciate structure and function.

References

Kapit, W. (2002). The Anatomy Coloring Book, 3rd edition, Daryl Fox.

KD Web Ltd (2011). The Meaning of Colours in Design. Available online at: www.kdweb.co.uk/blog/meaning-colours-design. (Accessed 15 June 2015.)

Wright, A. (2015). Psychological Properties of Colours. Available online at http://www.colour-affects.co.uk/psychological-properties-of-colours. (Accessed 15 June 2015.)

5 Coming Unstuck

> *The greater the obstacle, the more glory in overcoming it.*
>
> Moliere

The chapters of this book have to a large extent focused on development concepts for anatomical learning. Within this, there is an assumption that the student knows how to learn, and that learning is a natural process for knowledge development and exchange. Learning is a subjective experience that can be inspired by engagement with teachers as facilitators of knowledge. Each learner has the unique ability to learn, or choose not to. The barriers to learning create opportunities to consider what it is that obscures the passage to identifying truth in knowing and knowledge application. This chapter has been developed to unravel some of the obstructions that could occur when learning and/or teaching anatomical and physiological concepts. The chapter provides examples to illustrate how best we can build bridges between concepts and navigate the terrain of anatomy and physiology study.

What does it mean, when we believe something based on an authority? It means that we are taking something or someone else's words as truth, without a real knowledge ourselves. We all do this for many subjects. Our first authorities are the people who raise us. This is because we are born with no innate knowledge of the world, and have to learn it from scratch. To help us learn quickly, our brains are wired in childhood to largely believe without question what we are taught; we quickly absorb whatever our parents teach us.

We soon start learning from a range of sources, such as friends, teachers, books and other written material. As we learn and experience our world, we develop a map in our minds of what the world is like. This map becomes a truth filter. When we look at a new idea, we typically compare it to the mental map that we have. If the idea fits well in the map, we can add it. If the idea doesn't fit, we have a problem. We must either discard the idea, or make a change to the map. Change is difficult and often painful, so many people tend to discard ideas that don't fit their mental maps.

The goal of teaching is to show students how to find answers to troublesome learning content and move through a bottleneck situation. Pace (2015) describes the concepts of bottlenecks in decoding subject matter as an obstructed passage; one which may be difficult to bypass or navigate. You can almost imagine a bottleneck as a traffic jam, which you have to learn how to move through or find alternative routes around. Anatomy is a subject wherein a student may need to satisfy an examination as a required component of a module, or may really need to learn the content as a critical component for professional development and competency. Emotional bottlenecks such as increased anxiety can be created through avoidance strategies, lack of knowledge and inability to learn content. These bottlenecks, if not effectively managed, can further lead to heightened feelings of failure and inability to succeed.

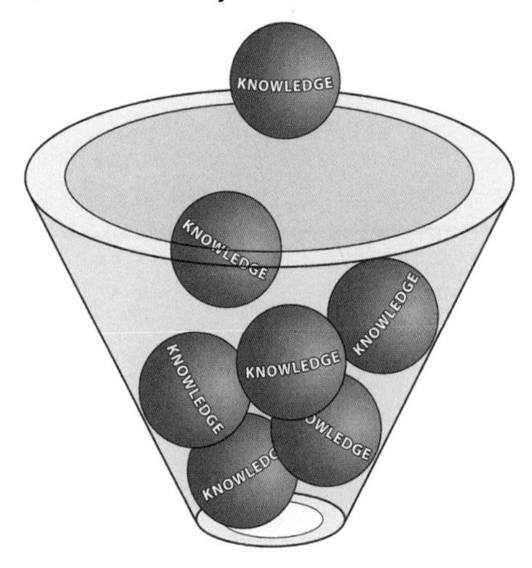

Figure 5.1: The concept of a bottleneck, whereby we learn how to move or navigate through an obstructed passage, where ideas or concepts may become stuck.

The Teaching and Learning of Anatomy and Physiology

Learning to Master Content and Content Application

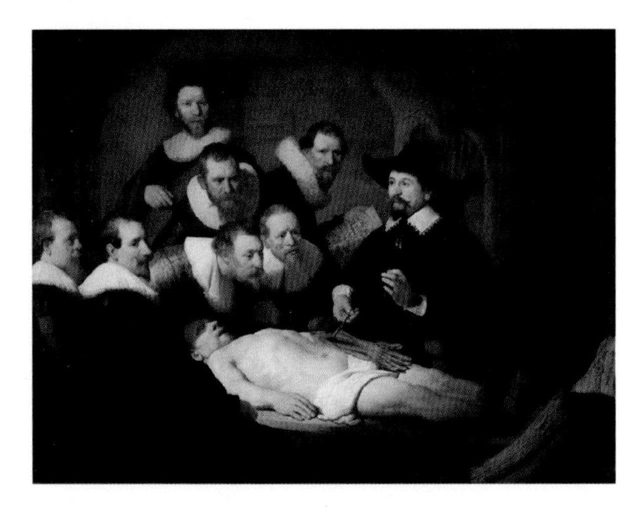

Figure 5.2: This painting by Rembrandt depicts the origins of teaching anatomy to medical students and doctors.

It is evident from the painting in Figure 5.2 that teaching is instructor led and students simply observe and hopefully learn what is being taught. Teaching, in this context, is more about turning individuals into scholars. The nature of teaching and learning has, to a large extent, focused on establishing truths about knowledge. In studying more advanced subject matter it is often difficult to appreciate what is or becomes important in terms of building knowledge. This could lead to uncertainty and doubt, which may impact the learner's self-esteem and ability to appreciate content.

As we teach we realise that learning may become complex and alien to the learner. The study of anatomy and physiology incorporates multiple bottleneck situations (alien spaces and places where learning is impacted), threshold concepts and troublesome knowledge.

To best illustrate the concept of troublesome knowledge and problematic learning, we have used art paintings as conceptual models, punctuating the content of this chapter. Artists have a unique ability to depict chaos and disturbance through their paintings by using a mix of colours and textures. The sea is used to denote turmoil and unexplored territories. The notion of riding out the storm to reach calmer waters is a common theme in art. Using this imagery, we can begin to consider the rough, or difficult, aspects of learning anatomy and physiology. The artwork in Figure 5.3 is an example of this imagery, wherein the use of colour and contrast creates an unsettled landscape filled with danger. The rocks symbolise the volume of risk.

Figure 5.3: Artists have a unique ability to depict chaos and disturbance through their paintings by using a mix of colours and textures. The sea is used to denote turmoil and unexplored territories, and the rocks to symbolise the volume of risk.

This chapter, unlike the other chapters, considers anatomy as a threshold concept, one that can create a troublesome state of knowledge and navigation of content. The reader will be taken on a journey into liminality – a transformative space where transformation of learning occurs and new knowledge or views of knowledge are formed.

A Stranger in a Strange Land

Have you ever travelled to a foreign land or place? What were your feelings, expectations, frustrations, highlights and disappointments? What happened when you arrived in a place that you had not previously been to, where the culture and customs are distinctly different from what you are used to, where language is a barrier not an enabler, where being different has huge implications for success and failure? The study of anatomy and physiology has the potential to be this place, often alien and isolating, using language that is different and processes that seem strange and obscure. How do we cope in such a space, how do we learn to navigate the complexities of the content we need to master, how do we learn to learn, and turn obstacles into opportunities? Imagine, if you will, that you need to ask directions in this strange place, yet your communication strategy is met with discontent and ignored – how do you advance? What happens if the advice you need is not the advice you get? How do we learn to make sense of perceived alien concepts and terms?

In considering the content for this chapter, I needed to reflect on my personal journey both as a student and, more recently, as an educator in anatomy and physiology, to identify spaces and places where learning anatomical content becomes problematic. There are many such places both within texts and in face-to-face learning situations. I recall one of my first learning experiences in my first year of anatomy. I was studying medical anatomy, learning the art of dissection. My lecturer was an experienced academic who took pride in instilling a sense of fear in his students, focusing on failure not success. I was scared to ask questions and simply nodded to indicate I understood, even if I did not, and prayed I would never be asked to evidence my knowledge, or lack thereof, in front of my classmates. My prayers were not answered and I was asked to describe and explain the muscle action of the upper limb in response to a handshake. I recall freezing almost like a deer in the headlights, and searched through the filing cabinets of my mind to produce some sort of educated response. My fear was palpable, with colleagues looking at me and almost acknowledging that this was how they felt as well. I now can appreciate that physiologically my autonomic nervous system was in overdrive, with my sympathetic nerves and parasympathetic nerves turned on full blast to create a frozen moment. I managed to stammer out a few muscle names and was asked to sit down. I realised that for my lecturer the joy was in watching me struggle. After my struggle and search for an answer, he lost interest and was not at all concerned with what I said. This lonely journey of searching for an answer can, at times, be troublesome, especially if the knowledge we have has no significance to the answer we need to produce. Students often get lost in the sea of anatomy, and feel the content and themes of the human body simply washing over them, without being able to absorb and enjoy the content sufficiently.

This feeling of drowning in knowledge has psychological implications for learning and knowing what to learn. How do we learn to dissect complex clusters of knowledge into smaller manageable learning sets? How do barriers to learning create opportunities to navigate knowledge and transform the way we think and understand? These questions are important in realising the complex nature of learning techniques we use to overcome hurdles. How useful are superficial answers and explanations? Does using technical terms out of context instil a sense of knowing, or expose the gap of ignorance? Are we simply imposters trying to get by without being detected?

In teaching anatomy and physiology, there is a natural assumption that learners will pick up on the main themes and fill in the gaps. As an educator I am often concerned about the impact on a learner's learning of leaving out important conceptual steps. Teaching may involve student instruction wherein crucial learning steps are omitted as the teacher assumes that the knowledge is already known by the student, e.g. it may be necessary to explain to a student that it is necessary to plug a computer into the wall socket before it will work. Think about the instructions you give and whether they are readily understood. Students may need to find missing pieces before they can learn to navigate difficult concepts. Some students are able to piece together knowledge despite not all the components of the knowledge being readily available or accessible. These learners may enjoy the challenge of building knowledge by forward thinking and conceptual inlay. Other students, however, require precise instruction in terms of understanding and then applying knowledge in a meaningful and directed way. Once the learner has mastered a concept there is a change in the way he or she thinks and, possibly, acts; simply put, the student has transformed his or her ways of knowing. Transformation is a rich concept signposted by change and conceptual mastery; it brings a sense of learning pride, but simultaneously propels the learner into a new matrix of learning development.

Transformative Learning

To fully understand the concepts of transformation and transformative learning, it is necessary to consider the process of transformation. Take, for example, a piece of food: it has a shape, a form and a specific texture in its raw state. If we ingest the food, we begin to change its texture, shape and form, which will be impossible to replace. As we continue to chew the food, we convert the food into a bolus. The process of chewing and digesting food, i.e. changing the dynamics and dimensions of the food, are akin to transformation and transformative learning. A trigger is needed to initiate the process; once transformed it is difficult to regain the original shape or make claims for unlearning. In other words, it is impossible for the bolus to return to its original raw shape. The food has crossed a threshold, a portal that has long-lasting effects on how it looks, tastes and feels. For the learner, this threshold is important in terms of recognising the impact of learning new skills, or digesting new knowledge.

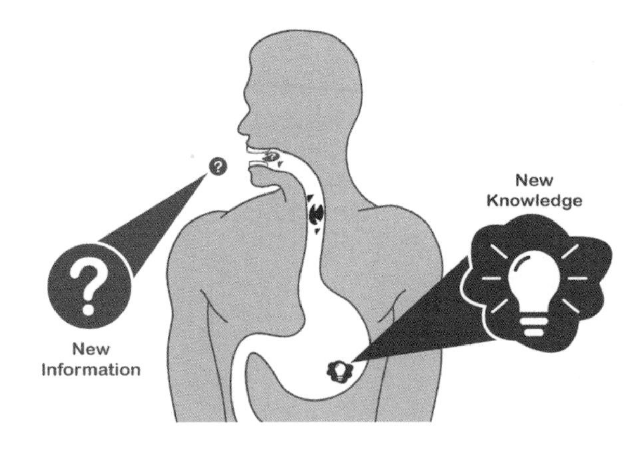

Figure 5.4: The process of chewing and digesting food is akin to transformation and transformative learning.

In the study of anatomy there are concepts and terms which are difficult to learn, conceptual nightmares which impede learning development. For example, students may be able to consider how a muscle works by visualising the shortening and lengthening of muscle fibres, but struggle to understand the chemical reactions necessary to produce such action. Students could spend significant time trying to conceptualise and draw meaning from diagrams and language they fail to understand and apply. These difficult areas in learning can be navigated or bypassed to avoid barriers and pitfalls. Take, for example, the driving of a car. The driver navigates a journey by choosing a course to travel. This course may include avoiding certain roads, areas or boundaries. The driver will eventually reach the destination. What happens if the areas that were avoided are now necessary inclusions in the journey itself, wherein the driver needs to pass through the avoided area instead of bypassing it? Passing through a previously avoided area could now bring a new perspective to the journey, reduce travel time, or may even introduce the driver to a new way of driving and planning future journeys. If we expand this example into the learning of anatomy and physiology, the same thinking could apply to learners who avoid learning and/or understanding difficult concepts. Such learners may eventually be able to pass an assessment without knowing but, likewise, by knowing, new opportunities to apply anatomical knowledge become evident and trigger stronger associations between concepts and processes.

Think Box

On a sheet of paper, write down any content from your anatomy learning that you chose to skip over. Think carefully about why you avoided learning the content. By avoiding the content was any of the content storyline lost?

Think about a film or book where you missed out a significant section. Could you follow the plot, or did you need to go back to make sense of relationships and associations?

In this book we have used the imagery of building a house to symbolise the study of anatomy. The key consideration in building a house is to ensure a solid foundation. If a house was not built on a solid foundation, would it stand? As a simple answer, it may well stand, but for how long? It is not a question of simply building knowledge or navigating around difficult concepts on an unstable foundation, but more a question of how long knowledge will hold before the cracks show, or before students are forced to face what they don't know or have failed to learn. Like a builder, we can learn to plaster over the cracks and patch up the foundations and brickwork, but the cracks are likely to reappear and could cause more structural damage. If we learn to open a crack, fill it from the inside and consider the structural repair, the final outcome may be longer lasting. In the study of anatomy we can brush over concepts superficially, but if questioned on a deeper level, the cracks or repair job within our knowledge may show through, creating doubt, embarrassment and possible disappointment (see the 'Jabberwocky' poem in Chapter 2 – Speak Anatomy). Do we simply need to know answers, or learn the process for establishing answers? If students are given answers, this may retard their ability to consider how those answers were established

The painting in Figure 5.5 depicts, in part, the safety within the womb. The womb is a peaceful, nurturing place wherein one feels comfortable. Outside the womb is uncharted space, the undiscovered, the perceived threatening world of the unknown or unexplored. Stepping out of the womb creates feelings of doubt, apprehension and anxiety. At times, stepping out of the comfort of the womb may be a powerful motivator for the discovery of new knowledge and new beginnings. The painting uses powerful images and concepts to communicate a message of safety versus unknowing, and holds unlocked potential for framing the study of anatomy within the concept of the unknown and the ability to discover what lies beneath.

Figure 5.5: 'The Garden of Eden', Folio 25v of Les Très Riches Heures du duc de Berry, a 15th-century illuminated book of hours held at the Musée Condé, Chantilly.

The human body, or study thereof, can be seen as a canvas within itself, enabling students and teachers to peel away the layers to study, examine and assess the content of what lies beneath. This exploration of the layers of learning within the human body reveals a tapestry of interrelated structures and mechanics primed for movement and function. The learning of how things work in relation to other structures creates a complex web of understanding which forms the building blocks of anatomical knowledge. The challenge is how to piece the different structures together so the end product makes sense.

The Challenge Within the Study of Anatomy and Physiology

Outward Facing In, or Inward Looking Out?

For many students, the ultimate challenge is learning an answer to a question. It is not so much the learning of the answer that may appear problematic, but the principles behind this type of learning. Students may need to be told what to learn in order to satisfy the requirements for an examination or assessment task. The focus is on learning to navigate the learning content and, more precisely, learning to leave out some of the content.

Extrinsic motivation applies to learners who are externally driven to achieve. Such a goal could be achieving a good grade in an assessment, as it may provide more meaningful rewards and opportunities in the future. This form of motivation/reward is important, but may also be damaging to personal development and knowledge evolution. If we simply learn to achieve external rewards, we may have lost the significance of the learning altogether.

Intrinsic motivation pertains to inner beliefs about learning and learning achievement. Students who exhibit this attribute tend to see learning as a personal journey, one which drives the love of learning, the ability to enquire, and seek out knowledge and truth about phenomena. Intrinsic motivation enables learners to develop new ways of learning, identify areas for further development and focus on how best to achieve learning success.

The challenge, therefore, is learning how to channel extrinsic motivation so that it enriches intrinsic inspiration, driving forward learning and learning change, and creating a nexus between knowing and knowing how. This transformative process may take time as some students may need guidance on how to develop internal learning attributes, reflect on experiences, and learn to develop multiple ways of doing and seeing anatomy from various perspectives. It is about moving into and through the conceptual transformative learning portal to discover new ideas, new strategies and new knowledge.

Liminality

Liminality is derived from the Latin word *limen*, meaning 'a threshold'. The concept of liminality is closely associated with threshold concepts and has an effect in teaching and learning. To some extent, learning can be seen as a ritual, a repetitive activity whereby learners dip in and out of learning resources to appreciate content and begin the process of content application.

Consider students in a library ahead of assessments. There appears to be a set time, space and ritualistic approach to how the revision is organised. Whilst in the library, students may decide to speak to other students. These interactions could instil a sense of fear in the student who has yet to learn or understand the knowledge content for the assessment. Once a student has mastered a concept, confidence in his or her answer and ability to engage in discussion with other students may become more meaningful. Here the student has been through a transformative process, the proverbial penny has dropped and the 'eureka' moment identified. The student is able to return to his or her revision and make sense of the learning context. In this new domain, there may well be further troublesome spaces and undoing of previously held beliefs and virtues. It is this constant settling into the unsettled that changes the 'learnscape' and propels the student towards a new, and hopefully deeper, level of learning and knowing.

Threshold Concepts

> *Here is Edward Bear, coming downstairs now, bump, bump, bump, on the back of his head, behind Christopher Robin. It is, as far as he knows, the only way of coming downstairs, but sometimes he feels that there really is another way, if only he could stop bumping for a moment and think of it.*
>
> A.A. Milne, 1926, *Winnie-the-Pooh*

The above quote challenges learning and focuses, to an extent, on new ways of thinking and doing. This idea is important for threshold concepts, in terms of realising and navigating them successfully.

Figure 5.6: Christopher Robin with Edward Bear.

A threshold concept is thus seen as something distinct within what teachers would typically describe as 'core concepts'. A core concept is a conceptual building block that progresses understanding of the subject; it has to be understood but it does not necessarily lead to a qualitatively different view of subject matter.

A threshold concept, by its nature, is likely to be:

1. *Transformative* – in that, once understood, its potential effect on student learning and behaviour is to occasion a significant shift in the perception of a subject, or part thereof

2. Probably *irreversible* – in that the change of perspective occasioned by acquisition of a threshold concept is unlikely to be forgotten, or will be unlearned only by considerable effort

3. *Integrative* – that is, it exposes the previously hidden interrelatedness of something. Often (though not necessarily always) *bounded*, in that any conceptual space will have terminal frontiers, bordering with thresholds into new conceptual areas.

Threshold concepts and bottlenecks, as earlier mentioned in this chapter, tend to be used interchangeably but have distinct and subtle differences. Anatomy and physiology present numerous applications that lead to transformation of learning and challenges within learning and mastery.

The table lists differences between threshold concepts and bottlenecks (from Pace (2015)).

Threshold Concepts	Bottlenecks
Small set of key concepts	Broader set of concepts that impact learning
Broadly transformative	Impact is not always global
Focus on what students know	Focus on what students do
Begin with subject consciousness	Begin with subject unconsciousness
Define a problem	Define a problem but simultaneously provide steps to address it
Example: failure to understand the mechanism by which urine is formed in the convoluted tubules	*Example*: not knowing the functions of the kidney
Example: being unable to give examples of hormones and enzymes produced by the pancreas, listing their functions	*Example*: being unable to differentiate between endocrine and exocrine organs

It is interesting to note that threshold concepts begin with a premise that the student has some subject knowledge and is aware that areas within the subject field are becoming or have become problematic. Bottlenecks tend to focus on unknowing and loss of knowing identity, but engage learners by considering fundamental learning steps to address inherent areas of subject difficulties.

Areas within anatomy and physiology that may give cause for confusion have been explored in Chapter 1 – Laying Foundations. There, we looked at the organisation of cells into tissues and organs, the overall concepts of internal support of the body, the mechanisms of how we move (locomotion) and how we know where our bodies are in space (proprioception), the integration of form with function, and finally the mechanisms by which fluid moves around our bodies. If these areas are not fully understood early in anatomical study, they may become bottlenecks (including emotional bottlenecks) to future learning, so take time and effort to embed the understanding and knowledge.

Navigating difficult learning content and spaces requires a mature appreciation of learning content and an acknowledgement by the student that there is a conceptual barrier to learning. The teacher may be able to play a role in helping the learner understand these conceptual barriers and implement learner outcomes to address imbalances in learning and knowing. One possible way in which students may be able to deal with these complex knowledge domains is by developing reflective learning skills. Reflective learning is discussed in Chapter 8 – Own Anatomy, on anatomical toolkits and reflection.

Is That Your Final Answer? The Role of Uncertainty in Learning

So far we have discussed difficulties in learning and navigating learning content. Learning can, equally, be uneventful, with the student appreciating the necessary processes and development steps within the learning cycle. Typically, learners new to the study of anatomy and physiology will not yet appreciate that they cannot know it all within their first year of learning. There is a sublayer of uncertainty that permeates the fabric of learning and provides a learning transition between entry and engagement of content. This uncertainty could negatively impact the student's emotional state and instil a sense of fear of the unknown and create a sense of frustration. It could equally motivate and provide a trigger for inquiry, propelling the learner into a cognitive search to seek out possible solutions to what may appear to be complex questions and problems. In my anatomy classes I ask my students to keep a 'bug book'. This book is their personal space to jot down any problems or issues they face within their study of anatomy and physiology. The book serves as a diary for concerns and provides students with an opportunity to acknowledge their uncertainty and work collaboratively with other learners to assess whether there are commonalities within the uncertainties they experience. It is interesting to witness the annotations within these books as students learn to juggle fear, pride, ability, competency, being, doing, asking and living the applied experience. Uncertainty is a useful trigger to encourage a shift in thinking and doing. Teachers would do well to signpost this to their students.

In learning, there is exploration of the unknown – the uncharted seas, the uncomfortable spaces and places wherein confusion, doubt and uncertainty reside. The ability to tolerate and recognise that uncertainty is part of learning, and that learning leads to uncertainty, is a critical process of learner maturation and personal development.

Difficult Concepts in Practice

'The more I study, the less I learn,' is a complaint of many a student. This statement epitomises the point of knowledge and/or information saturation. Imagine a child trying to stuff his mouth with as many sweets as possible. The more the child tries to cram sweets into his mouth the less he is able to digest them. In anatomy, knowledge or information overload occurs through bottleneck situations. Students typically try to learn too much without fully digesting and appreciating what it is they need to learn. The key to learning is to do it slowly and progressively, being able to digest small chunks of knowledge before creating room for more. If the knowledge we have is the cause of the bottleneck or jam, then that jam situation needs to be cleared before more knowledge or information can enter.

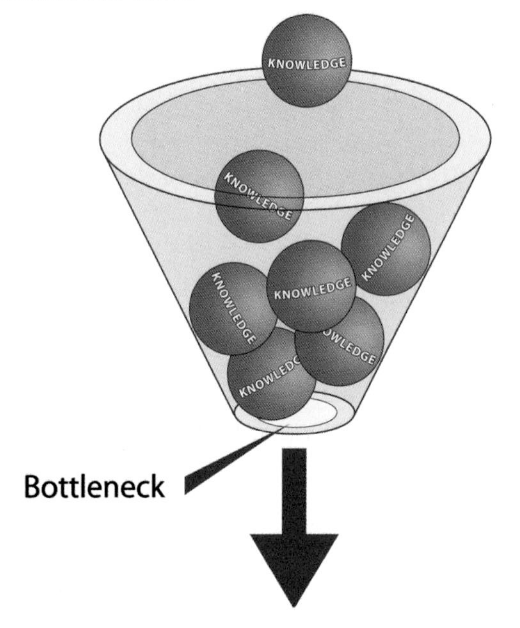

Figure 5.7: The student tries to assimilate too much information, causing a bottleneck or jam, which needs to be cleared before taking in more information.

This concept of ingesting and digesting is useful in paving the way for learning and navigating difficult concepts. Students need to learn that when nothing seems to be going in, there is probably too much inside that needs to come out first. Just as the child who has a bellyful of sweets needs to digest and absorb the sugar before he is able to eat his dinner.

How do we help students learn and navigate difficult learning concepts and terrains?

Learners may often unknowingly get the subject matter wrong. It is important for teachers to weigh up a delicate balance of what students need to learn, as opposed to what students need to do to become successful in a discipline and specific learning area. The cartoon in Figure 5.8 provides an amusing illustration of this point.

Figure 5.8: Learning as an active process.

Teaching in anatomy and physiology involves making leaps in learning from both a cultural and an emotional perspective.

How do students come to understand the ways of operating that will enable them to respond to the intellectual challenges of learning anatomy?

Some students are able to pick up on concepts and accelerate their learning and application of knowledge, whilst others need time and direction to understand and appreciate the relevance of the learning content. Often, learners who fail to realise how to learn and work through difficult concepts exhibit a fixed mindset. These learners may be comfortable not knowing and be happy with what they know. Students who are keen to develop further understanding and integrate their knowledge tend to exhibit a growth mindset, one which enables them to complete the missing steps and better understand concepts, applications and complex processes.

The conceptual illustration in Figure 5.9 tries to position the learner within a connected framework of the known, that still to know and learn, and the unknown. The concentric circles depict a ripple effect of knowledge being circular and moving through different ripples to acquire and apply new knowledge. Students may be satisfied with staying in the inner circle and focusing almost exclusively on what they know. Educators need to encourage movement between circles, so that knowing can be assessed and the unknown eventually becomes the known as new knowledge is formed and secured.

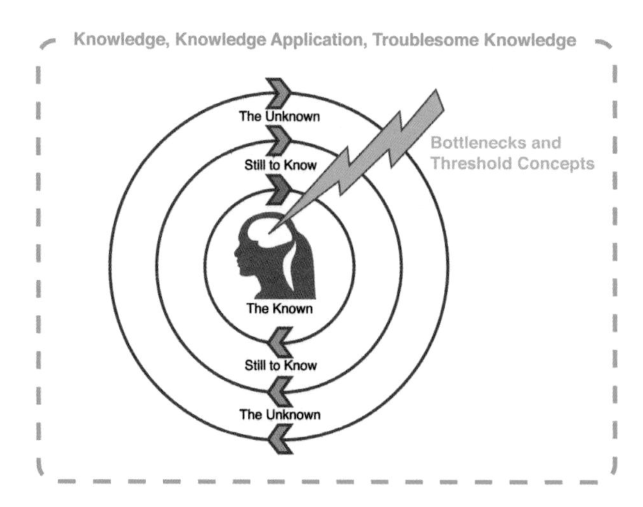

Figure 5.9: Knowledge circles – the relationship between knowledge and knowing.

Students need to be shown the steps for success. This involves teaching students how to ask basic questions and contemplate procedural outcomes and challenges.

What are the steps that experts take automatically, that students need to learn in order to better understand anatomy as a field of study? How can the teacher facilitate the learning of difficult concepts in anatomy and physiology?

The Steps in Decoding or Navigating Difficult Learning Concepts

(Adapted from Pace (2015))

- *Define a bottleneck or area of problematic learning.*

  Identify a place in a course where many students encounter obstacles to mastering the content. This could be system specific, or general anatomical terms.

- *Define the basic learning tasks.*

  Explore in depth the steps that a more experienced student or expert in the field would go through to accomplish the tasks identified as a troublesome learning concept.

- *Model these tasks explicitly.*

  Let the student observe the teacher going through the steps that an expert would complete to accomplish these tasks. This could be achieved by providing a metaphor or analogy for the desired thinking, e.g. like water through a strainer to illustrate how fluids move through semipermeable structures, or performing the desired thinking in front of students with a subject-specific example, or explicitly highlighting crucial operations in the example. This would involve working carefully through each step of a process, explaining how factors work to produce an effect, e.g. decoding the sliding-filament theory of muscular contraction. It is important for learners to understand expertise and expert knowledge. Experts may be experienced within the subject matter and have mastered skills that enable them to reason and locate information in a systematic fashion. An expert is not someone who professes to know it all – quite the opposite. An expert has learned how to know and has developed skills to identify what it is that is unknown and needs to be learned. An expert is able to direct focus within these areas of learning.

- *Give students practice feedback.*

  Construct assignments, team activities and other learning exercises that allow students to do each of the basic tasks defined above and get feedback on their mastery of that skill. It is important that feedback is used *for* learning and is not simply a marker of learning.

- *Motivate your students.*

  Decide which approaches encourage students to excel and then utilise them to create or define an environment that fosters a positive learning environment. It is important to simultaneously identify any emotional bottlenecks that could impede student success and that could arise from learners' preconceptions of the field of the material being studied.

> *The mediocre teacher tells. The good teacher explains. The superior teacher demonstrates. The great teacher inspires.*
>
> William Arthur Ward

- *Assess how well learners are mastering these learning tasks.*

  Create forms of assessment that provide specific information about the extent to which students have mastered the particular learning tasks identified in the second step above, *Define the basic learning tasks*. It is important that assessment addresses individual learning development and mastery of the task as opposed to general nonspecific assessment techniques.

Figure 5.10: Assess how well students are mastering the learning tasks.

- *Milestones and markers.*

  Celebrate when your students have mastered the task or process. Help them record their success and enable them to monitor their progress towards fully realising how to apply their learning.

- *Master mentorship.*

  - Help students see what is needed to become confident and capable in task execution and knowledge generation. Mentorship is about facilitating learning by enabling learners to develop skills of reflective practice, problem solving and knowledge construction. The key to positive and successful mentorship lies in the mentor's ability to know the student. Mentors could:
  - Encourage discursive learning by giving the student a voice that is listened to.
  - Put content into a shape or form and support student learning by building capacity for autonomy.
  - Help learners change expectations.

- *Start big.*

  Before students can begin to appreciate the smaller details, they need to have mastered the big ideas and concepts. It is important that these big ideas and concepts are fully explained and that learners have learning opportunities to master them, before beginning to dissect the finer details. For example, before teaching the intricate processes within the sliding-filament theory, students may need to see a muscle working and conceptualise this working first, or consider the digestive system components before learning the process for each component.

- *Share what you have learned about your students' learning.*

  Establish forums to share what you have learned about your students. Dissemination of learned information from student learning and engagement will enable programmes in anatomy and physiology to consider key learning activities to enhance learner success.

Conceptually, these steps involve intricate feedback and feed-forward loops, traversing a highway of interconnectivity. They are not static and certainly do not operate in isolation. The interrelatedness of the steps creates a spiral that evokes a learning cycle of mindfulness, mastery, metaphors, mistakes and modelling.

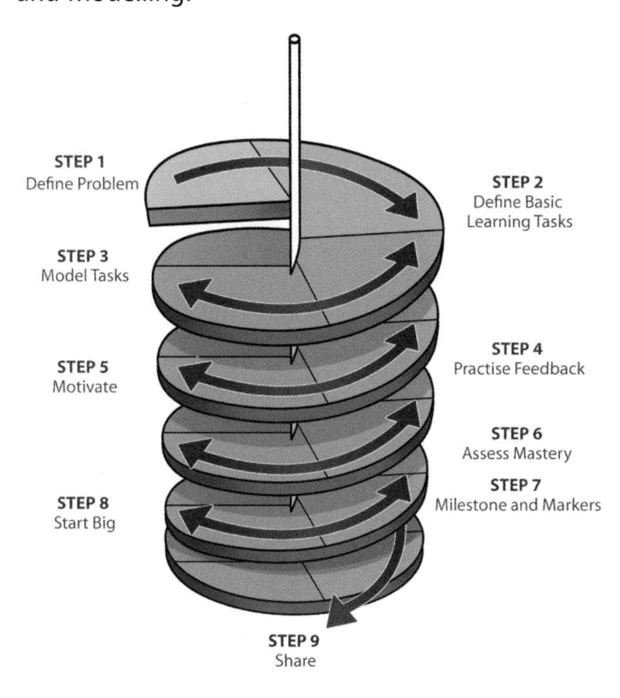

Figure 5.11: Spiral diagram of the steps in decoding or navigating difficult learning concepts.

Decoding Anatomy

Teachers and educators can further use strategies for making implicit anatomical and physiological subject matter steps explicit. These may include:

1. *Decoding interviews*

   Working with students and experts to dissect the core thinking behind the concepts. This enables the learner and teacher to carefully decipher and understand how individuals shape their learning by walking through the learning process.

2. *Metaphors and rubrics*

   Work with students to provide examples of how the process may appear.

   a. Metaphors create opportunities to align processes with practice ideas and better conceptualise complex learning.

   b. Rubrics may further enhance understanding by considering the criteria against which assessment judgements are made. This provides a framework for learning and creates a spreadsheet of descriptors to illustrate specific and acquired knowledge necessary for success and mastery.

3. *Systematic questioning*

Help learners use evidence systematically to support an interpretation or understanding of anatomical fact. This could include a series of structured questions that encourage students to realise the importance within and about specific concepts, processes and applications such as:

a. The questions that are posed to develop an interpretation.

b. The interpretation that is being defended.

c. The evidence used to defend the interpretation.

d. An alternative interpretation that would be shown to be less convincing than the initial interpretation.

Final Thoughts

Anatomy and physiology are a lived experience, wherein one needs to appreciate the workings and failures of the body and how structures relate, connect, travel and exit at critical points within the body. This allows students to consider truths behind knowledge, to ask different questions, and use evidence to support positions and arguments. Through questioning, a trigger for learning may be established to further enhance the learning approach.

Anatomy and physiology can be a puzzle, a difficult conceptual space wherein understanding and application can appear foreign, obscure and uncomfortable positions to be in. The learning is about knowing how to master smaller components and making internal changes to the way learners think, and the way teachers teach. Differently put, anatomical learning is about embracing and realising the power of change in both knowledge and individual development. It is about confronting change and learning to navigate new territory by connecting what we know with what we need to learn.

> *A bend in the road is not the end of the road unless we fail to make the turn.*
>
> Helen Keller

Reference

Pace, D. (2015). Decoding the Discipline Workshop. EuroSoTL, Cork, Ireland, 8–9 July 2015.

Learn Anatomy

> *I was never very good at exams, having a poor memory and finding the examination process rather artificial, and there never seemed to be enough time to follow up things that really interested me.*
>
> Sir Paul Nurse

If a Nobel Prize winner and renowned cancer scientist feels like this, it is no surprise that most people hate exams. How do you feel about them? Have a think …

> *It's the day of the written anatomy and physiology examination at last. Take a few minutes to imagine it: visualise yourself walking into the examination hall … sitting waiting for the invigilator to tell you the examination is about to start … reading the first question … holding your pen … How does it make you feel? Terrified? Exhausted from a couple of all-nighters? Sweating profusely and heart racing with fear? You are so nervous that all knowledge of the digestive system has disappeared, and the only thing you know about your stomach is that it is churning in terror?*

And that is just how the teachers feel! Yes, teachers worry about exams too! They frantically pace the floor outside the examination hall like a mother hen whilst their little chicks are scribbling away.

It may seem incredibly premature to confront the examination preparation before the course begins, but it truly isn't. The vast majority of anatomy and physiology students are learning the subjects as stepping stones to qualifying in a different career, whether that is in biomedical science, medicine, holistic and beauty therapies, or sport and exercise sciences. Even though the subject matter in the anatomy and physiology class may be very interesting, for most of the students, the examination will appear to be the most important part of the course.

This chapter aims to help students develop study skills in order to prepare for examinations, including how to manage the essential revision required. Learning key study skills will also firmly embed the anatomy and physiology knowledge rather than just learning to pass an examination, giving a long-lasting memory of the facts and concepts. Teachers, do carry on reading, as by following some of our tips, you can help your students best prepare for their upcoming examinations. The key word here is *preparation*. This means right from day one or before day one, rather than two days before the exam. So let's start now!

> *Before anything else, preparation is the key to success.*
>
> Alexander Graham Bell

Preparation Rather Than Procrastination

Recognising what you don't know is key to targeting revision effectively. Spend some time considering the categories in the table.

Category	What it means	How you find out
Known unknowns	All the things you know that you don't know	Intended learning outcomes Past papers Reading homework comments VARK assessment Examiners' reports Monitor your procrastination activities carefully Draw up a revision plan Work with a study buddy
Unknown unknowns	All the things you don't know that you don't know	
Errors	All the things you think you know but don't	
Unknown knowns	All the things you don't know you know	
Taboos	Dangerous knowledge	
Denials	All the things you are avoiding	

Once you have decided to knuckle down and work at preparing for your assessment, be aware of any self-sabotaging activities and avoidance strategies that you may be engaged in. The evasion of doing something which must be done can result in feelings of guilt and stress that compound and build, resulting in yet more avoidance and procrastination. Excessive procrastination will have a negative effect on studies, so it's important to be able to identify procrastination and recognise how to stop it.

Why Do Students Procrastinate?

- *Feeling overwhelmed*: A sudden onslaught of deadlines combined with poor time management can distort your sense of time, and without prioritising your revision, it may feel impossible. This in turn may create feelings of anxiety and stress, which will hinder the revision process.
- *Task difficulty*: You may be unsure how to start revising. This can be a common problem for anatomy and physiology students, as it involves mastering knowledge and concepts that may be new and challenging.
- *Disliking the task*: Many students procrastinate if they find their work boring or unpleasant, particularly if they are attempting to memorise dry facts.
- *Perfectionism*: Students can set unrealistically high standards for themselves. Fear of not achieving these standards can cause procrastination.
- *Poor concentration*: A work environment full of distractions, noise and mess can make it difficult to focus on the task at hand.
- *Lack of motivation*: All of the above can make students feel less motivated about completing their work and revision; therefore, they are less likely to be self-motivated.

(From University of Bath (2015a))

Signs of Procrastination

It is common for people to mask procrastination by occupying themselves with tasks unrelated to the object at hand. Cleaning the recycling bins, washing the kitchen floor and tidying the desk are all tasks that suddenly seem far more important than anatomy and physiology revision or homework.

The following statements, if true for you, are signs of procrastination (from University of Bath (2015a)):

- I delay starting a task because I find it difficult.
- My mind wanders to things other than the task at hand.
- I sometimes wonder why a task needs to be completed.
- I often give up on tasks once they start to become tricky.
- I have difficulty getting started on a task.
- I have so many things to do that I'm never sure where to start, so hardly any of them ever get done.
- I often put off a task in which I have little or no interest.
- I will ignore a task when I'm not sure about how to start or finish it.
- I often start a task without finishing it.
- Frequently, I can't decide which of my tasks I should tackle first.
- I find myself ignoring a task, hoping it will go away.
- I choose to do the housework rather than study even though my house is clean.
- I spend more time on social media than I do studying for my examination.

What Can I Do About Excessive Procrastination?

Along with understanding how it affects you personally and emotionally, better organisational skills and time management are the key to defeating procrastination. Here are some ideas which may help you to control procrastination:

- Improve your time-management and organisational skills (see University of Bath (2015b)). This includes prioritising your workload and working out a plan of action.
- Commit to completing tasks once you have started them. Give yourself a reward when you complete a section to keep your motivation levels high. Treat yourself to a cuppa or a magazine, and don't let yourself have it until you finish your intended task.
- Break large projects into small, manageable parts. Attempting to tackle huge topics is demoralising.
- If appropriate, don't do it alone! Find a study buddy to work with; don't forget you can link to fellow students via Skype, text or email. If you are struggling with your work, try to discuss it with your lecturers or peers.
- Find a good place to study or work, either in your own home, or at a library or even a coffee shop. We are all unique in our needs; some students need a background hum of hustle and bustle as a form of white noise, whereas others need absolute silence. Work out which you need (rather than which you prefer). Minimise distractions and maximise your comfort. Quiet music in the background if it helps you. Turn emails and social media OFF!! Have a drink and a snack to hand at the start of your study period, and do not move from the task until you complete your intended session. Reward yourself with another drink.
- Try not to be unrealistic about what you can accomplish. Set yourself standards which you can reasonably meet and be happy with.

- Boost your motivation by reminding yourself of your strengths and tasks you have already accomplished. Think about making yourself a star chart. It may seem more suitable for a child, but rewarding yourself with a sticker on a chart is very self-satisfying.

- If you get stuck with your revision, try a different study strategy. This could mean switching your method of study, or working on a different area of the same task. Do not just put the books down and ignore the issue.

- Discover why you are procrastinating. This may make it easier to overcome it. You may be able to help yourself but, if not, you could consult a tutor, supervisor, friend, study buddy or counsellor for help.

Above all, think positively and get going. You will feel increasingly relieved as you get closer to completing the task, and you will come to look forward to the feeling of satisfaction experienced when you have completed it. Imagine that the revision is a similar process to running a marathon. The end goal of the examination is like the actual race. In order to successfully complete the mammoth task of running just over 26 miles, athletes must be self-disciplined in starting the preparation by choosing a start date to begin the training process. Little and often, the athletes run, increasing the distance slowly but surely until they achieve the required distance in a competitive time. If athletes do not train their bodies by this preparatory training process, they will surely run out of steam in the marathon itself, and will not achieve their ultimate goal. Students must prepare and revise sufficiently early to allow themselves to complete all of the planned revision in time for the examination. Insufficient revision in insufficient time will only lead to insufficient grades.

Make OSCAR Your New Best Friend!

O	Organisation
S	Selection
C	Creativity
A	Association and asking
R	Repetition

Figure 6.1: OSCAR.

OSCAR is a handy mnemonic for a strategy: *organisation*, *selection*, *creativity*, *association* and *repetition*. Taking each in turn, we offer you some new strategies for learning with tips to give you a unique 'learnscape' for your studies and create a pathway to success.

Organisation

Confidence in examinations comes from being prepared and organised, and in that preparation becoming familiar with the examination format and structure so there are no surprises. Revising and retaining the information are of course still incredibly important, but can be structured such that the knowledge is packaged or presented in ways that suit the examination.

Sort through your notes before you begin and organise them into a sensible structure. Start by sorting your resources, relevant books and classroom notes. Get rid of scrappy bits of paper. File everything that could be useful to you and slot it into the relevant section. The reality is that you *must* make time to avoid panic, which is debilitating and totally counterproductive, so get organising now.

Diarise Your Revision

Rather than spending hours making a pretty revision timetable that takes longer to make than it does to revise, use a diary to record all activities that you already have planned, and all appointments, lectures and compulsory activities. Include absolutely everything that isn't revision. This creates a diary that tells you when you can't revise, making it very clear when you do have free time for revision. Now block out that time as revision. Make it a fixture that is not replaced by any other task.

Intended Learning Outcomes

The first important task for organising revision is working out what you already know, what you don't know, and what you need to know. If you have not been provided with intended learning outcomes, ask your tutor. Identify the topics and areas which you categorically do not remember and highlight them.

Past Papers

After looking through the intended learning outcomes, the second port of call in examination preparation has to be the past papers. Students should be given access to these early on in the course. If you have not been given them, then ask for them. Teachers, make sure you provide them.

Reviewing the structure of past examinations will mean that it is easier to prepare and the format will be familiar. Work through the past examination questions to ensure that the intended learning outcomes are met. Experiment with the questions so that you can substitute in alternative topics whilst maintaining the format. Viva voce examinations should also be practised, and model questions can be obtained from the examining body.

Example:

List the origin, insertion and action of the quadriceps muscle group.

Change the name of the muscle to any of those on the syllabus and you have yourself some alternative questions.

Compare and contrast the structure and function of the stomach and gall bladder.

This could be changed so the organs are substituted for any other organs within a system:

- Liver/pancreas
- Thyroid/parathyroid
- Pituitary/hypothalamus
- Anterior pituitary/posterior pituitary
- Short bone/long bone
- Myelinated/non-myelinated nerve, and so on.

Get creative!

Review Homework Questions

Homework topics are designed to be continually assessing the acquired knowledge. Reviewing the questions, and getting creative with them as well, will give a plethora of possible examination questions for both student and teacher.

Homework questions are doubly useful as examination practice, as the teacher will have already given guidance on missing information (feedforward), and will have given marks for the answer. Even if the question was poorly answered, there will be feedback to help the student frame the perfect answer – and with the wonder of hindsight, the perfect answer is usually more attainable the second time around.

Examiners' Reports

Learn from others' mistakes, not your own! Most examining boards produce examiners' reports, which give feedback on the quality of answers given at previous exams. These have a wealth of information, particularly on common mistakes that have been previously made. Forewarned is forearmed!

VARK Self-Assessment

Is your learning preference visual or auditory? Do you work better by reading rather than acting out the topic? Work out your learning style to maximise your revision by using the VARK assessment described in Chapter 3 – Do Anatomy and contained in the Appendix – The VARK Questionnaire.

Selection

Choose which topics you need to study and work out when to start them. Consider a diary sheet annotated with all the different topics you need to cover and tick them off as you study them. Start by studying a topic you find difficult or don't know enough about, so the information can 'settle' a little in your head. Remember how long it takes to learn something. Some subjects need 'percolation time'.

Vary the method of study; use websites, quizzes, and apps, as well as good old pen and paper.

Lists, Charts and Notes

Make a list of the topics that you already know well, and leave these until near the end of your revision. It is easy to feel very pleased with yourself when you revise a topic that you know inside out and back to front, but it is a procrastination exercise.

The conventional way of revising is to make lists of information, and it may well be that your brain likes this better than any other way. It is certainly the way that most people will tell you to revise, but you have been warned – your brain and their brains are different: find your own way and don't attempt to use other peoples' lists. Write them in a style that pleases *your* eyes. Do use colour to differentiate the lists, and use subheadings or bullet points to break down the information into manageable chunks. Chapter 4 – See Anatomy explains the use and impact of colour on learning.

If you want to revise in this way, and you are going to make lists, try to find ways of making them interesting and keeping them short. The temptation is always to write down too much so you are attempting to learn huge paragraphs. The production of 'shrinking' lists is a better option. These start off as a paragraph, which you condense over and over, so it grows ever shorter as you become more confident that certain areas of information are already in your head.

Remember, one of the most common mistakes of revision is that students tend to spend too long revising what they already know. Do not spend hours trying to memorise a list that you already know word perfectly, however comfortable you feel learning it.

Use Colour Where Possible

Chapter 4 – See Anatomy made you aware of the visual impact of the use of colour. Make your revision notes interesting to the eye and memorable. Use specific colours for specific topics. Highlighter pens, coloured pens and coloured sticky notes can all be used. Arteries in red, veins in blue and lymphatics in green are the usual colours to be seen in a textbook, but go further, with the digestive system in brown, urinary system in yellow, skin in orange, nervous system in purple and immune system in black.

Why not try underlining or colour coding particular pieces of information? A series of different coloured highlighter pens might be useful here. This is a very useful first step to breaking down long pieces of writing into more usable short lists or diagrams.

Make a Mind Map or Shrinking Mind Map

The concept of mind maps was also introduced in Chapter 4 – See Anatomy, and should be considered an effective revision strategy. The aim of the game in revision is to reduce a lot of material into a small space in order to simplify and clarify the information that is needed to be retained and understood. If you have produced a mind map, the idea of a shrinking mind map is to create the same chart with fewer branches. You have the same number of major 'legs' and use the same colours, but you do not write down all the detail. You might shrink the map once, twice or even three times, so that the final shrunken map might only have ten or twelve words on it, but each word will trigger a memory of all the other things that were on the original leg. This prevents you from wasting time revising information that you already know like the back of your hand.

Creativity

How do you want to see the information? Where do you study best? What things help you learn most? If you have completed the VARK assessment mentioned in Chapter 3 – Do Anatomy and contained in the Appendix – The VARK Questionnaire you will have identified your learning style and will have read up on the key tools that can be used to help you learn the best ... Now is the chance to flex those creative muscles and experiment with some different ways of learning.

Songs, Rhymes, Mnemonics and Acronyms

Cheesy but effective ways of getting you to remember key points.

The title of this suggestion may sound more difficult than the examinations themselves, but these can be fun and memorable ways of learning screeds of facts. Making up catch phrases or rhymes can help you with crucial bits of information. It may make you cringe but you won't forget it. The Amatsu Training School students of the class of 2010 will never forget the name of the fifth cranial nerve after their rendition of 'The 12 Cranial Nerves' song, set to the tune of *The 12 Days of Christmas*.

Mnemonics and acronyms can do a great deal more for you with less risk of procrastination. A mnemonic is a word or abbreviation that helps you remember. An acronym is a word made up using the first letters of a series of other words or the first word of a series of sentences. For example, to remember the order of the carpals, which is the more memorable list?

Example:

Can you remember this:

Scaphoid, lunate, triquetrum, pisiform, trapezium, trapezoid, capitate, hamate?

Or the memorable mnemonic:

Some Lovers Try Positions That They Can't Handle?

Flash Cards and Posters

As we saw in Chapter 4 – See Anatomy, flash cards can be a big help in remembering important information. Why not try making some brightly coloured lists or even just writing down key words that you want to remember and putting them up in your room at home?

Try different colours for different subjects or different areas of the room. If you can persuade your family to go along with this, another way of utilising this technique is by using different rooms in the house for different topics. If the dining room becomes *muscles* and the kitchen becomes *bones*, then moving from room to room can, quite literally, open up different 'files' in your head. Put *kidney* and *large intestine* information on the toilet wall and *stomach* information on the fridge or food cupboards.

In Chapter 4 – See Anatomy, in the section on visual learning, we introduced the concept of imagery. If you regularly use the location of the theory examination for your usual anatomy and physiology classes, you can assign different parts of the room to specific topics and either place posters in different corners or imagine them in position. Just glancing at the different corners of the room when in your examination will help your recall. Or close your eyes to visualise the rooms at home where you have stuck posters and diagrams. It is an unusual approach, but works for many people.

Models, Toys and Games

Make sure you read Chapter 9 – Resource Anatomy for ideas on using games, toys and models to study. If you have children, using games is a great way to help involve your offspring to get you to revise. Can you make a game that they might want to join in with? Children are like little absorbent sponges that could shame you into knowing your anatomy to keep up with them!

Association and Asking

Making links between the information will embed the learning and make it memorable and extend your knowledge. Adding notes and reviewing the material within your existing class notes and resources helps you do this.

If you truly are confused, then ask for help. Discuss it with a colleague rather than searching the internet, which can give conflicting information. If you still have an element of confusion then ask a teacher, but try to understand it yourself first. Get together with other classmates, and have regular study sessions, either face to face or via Skype or email.

Study Buddies and Study Sessions

Look around the anatomy classroom and work out who will be your 'study buddies'. You will quickly work out who you want to sit next to, or who is like-minded. Who do you respect in your class? Even if your course is an e-learning course, there is often a forum or Facebook group for the class. Your study buddy has the potential to be your new best friend, your surrogate family and an all-round pillar of support. It is good to have several study buddies if you have the opportunity; one who is on your wavelength so you can support each other and one who is on a different wavelength but who respects elements you can do too. Great exchanges of information can happen when you have a range of skills and talents to pool.

Arrange to meet up face to face or by email, Skype or phone. Swap ideas. Share thoughts about the class. Work together to plan homework and to work through past papers. Test each other on the lists and tables you have worked on together. Send out a group email with the Origin and Insertion of the Day, or Hormone of the Week.

Checklist for Study Buddy Sessions

Write a Lesson Plan

Sharing the learning is one of the most effective ways to learn. This is why your teachers are so amazingly knowledgeable! But good teachers use a good plan that identifies the learning needs of the group and the intended learning outcomes, and you will need to do this too, albeit quite informally. Taking the time to plan the session in the form of a lesson plan will be incredibly worthwhile. And it will be hugely empowering to share anatomy and physiology with your peers.

Picking a topic and sticking to it is one of the hardest tasks for a study group. It is very easy to go 'off-piste' and slide into other areas. By committing to a designated topic, and picking aims and objectives of the session, everyone can contribute to the intended learning outcomes.

Include some time at the beginning of the session for catching up and general chit-chat – half an hour is plenty. End with a general chat session too. Do not assume that everyone will walk in to the session ready to think *anatomy*. They will need to be warmed up by some general chit-chat.

Make sure you study your chosen topic in advance and bring questions with you about that topic. You will reap the rewards of your preparation.

Pick a Leader

In a group scenario, there is often a natural leader who regularly chooses to lead the session. However, even if there are just two of you in your study group, take it in turns to lead the session.

The leader will call the start of the session, keep it to task, keep chit-chat to the beginning and end of the session, and record any problems that need to be brought to the attention of the course tutor.

Keep it Simple

Set a manageable timetable. If you have a big task, break it down into chunks so you don't put yourselves off. Rather than attempting to revise the entire digestion system in an evening, break it down into sections like *ingestion*, *digestion*, *absorption* and *elimination*, and spend a session in detail on each of them.

Keep a checklist of topics that confuse you. If you can't work something out within your group, jot it down to discuss with your teacher or learning provider, and call a halt to the discussion rather than confuse yourselves further.

Be Selfish

If it doesn't work with your first choice of study buddy, try somebody else. This is your time to learn, as well as to enable others to learn, but YOU come first.

Listen, cogitate, digest and enjoy the sharing, but remember you don't have to think in the way that your colleagues do. If it doesn't work for you, dump it, and find your own way of learning! It doesn't mean one person is right and one is wrong, simply different in learning styles.

In a group scenario where time is limited, have a timer and allocate the time for each person to talk, question, or explain what they want to know, answer or prove their knowledge in. The time limit holds back the interrupters and the people who like to vocalise their thoughts, and gives the floor to the questioner.

Be Positive

Avoid the urge to discuss how hard the homework was, or how difficult it is to remember the functions of skin. However wonderful it feels to trade tales of misery and woe, it won't help your learning. Instead, remain upbeat and positive.

Repetition

Having sorted your notes, made time on a regular basis to study (in short blocks of time) and created your learning aids, the trick to memorising the information is to repeatedly revisit it. The first time you work through it, only a small proportion will be retained. Yet, the more times you revisit, the more information is memorised. The minimum of repetitions and rereading that is needed is three times of working through a topic, ideally more.

Find times in your daily life when you can repeat your anatomy and physiology learning over and over again. We introduced the need to speak anatomical language in order to make it memorable in Chapter 2 – Speak Anatomy. Talk to your colleagues and friends about your studying and involve them in your learning. Sit in the car and recite things from crib cards whilst waiting to collect the children. Standing at the side of the football pitch or sitting at the swimming pool waiting for your children is an ideal time to recite things or list items in your head on the topic of the day. Talk out loud to the cat/dog/goldfish/hamster and tell them all about bone structure!

Of course, you may prefer to ask a family member or a friend to test you from your coursework. Be careful though, as *Mastermind*-style question and answering does not necessarily guarantee that you will be able to apply and remember lists of information in an examination. You are probably better off being tested on the charts and lists that you are trying to remember for the exams than by having somebody look through your folder of work and asking you questions here and there. A good plan would be to build in a 'testing time' of a few minutes each day to see how well your revision is going.

Audio Format

Take some of the lists or the notes that you have made and record them in an audio format. *Variety* is the key. You might want to get different friends or members of your family to read different things on to the recording so a different voice will jog different memories. Other people will choose background music to match a particular subject or topic so that, in an examination, thinking of a particular piece of music will bring back the information you require.

Several good basic anatomy books have an audio format, so you can utilise this by listening to it whilst dog-walking, waiting for children or gardening. The great advantage of this is that you use it during 'dead time'. You would be unlikely to take your revision file out to the park when you walk the dog but there is no reason why you can't take your iPod or phone with you and listen. Don't forget what we have already said about controlling time. If you put together all the little pieces of dead time in a day you could easily find yourself between 30 and 60 minutes of 'listening time' every day. Just think how much difference that could make to examination success.

> *You have to learn the rules of the game. And then you have to play better than anyone else.*
>
> Albert Einstein

Preparing for the Examination

Practising Previous Examination Questions

> *Try not. Do or do not, there is no try.*
>
> Yoda

All examinations are written in a coded language because, to be honest, there are often not many different questions that can be asked about defined areas of anatomy and physiology. The same questions are asked in different ways, or wrapped up in what can appear to be confusing language. A key to success in examinations is *understanding* the questions that are being asked. This may sound obvious, but so many people miss out passing their anatomy and physiology examinations every year because, although they were well prepared, they have misunderstood the question that is being asked and have written down information for which they cannot be given marks.

Over 60% of all errors in exams are caused by not reading the question properly, so practising the examination styles used over the years will help you get used to the way the questions have been written.

Look through the past papers and work out the format so you are familiar with the layout of the examination. Are there some compulsory questions? Will you be able to choose from a selection of questions? Perhaps divide the past papers up, so you can practise a few questions each day, keeping one or two for a 'run-through' to see how you cope working to a time schedule.

There is a definite format in how the examination questions have been set. The whole syllabus will be addressed in some form, so you will have to revise every chapter and every body system. There will be some questions which are easy to answer, and some that really do test the information that lurks in the dusty corners of the textbooks. These 'dusty corner' questions are called 'extension questions', and aim to test students working at distinction.

You may be asking yourself, 'Can I write fast enough for long enough? What if my hand falls off in the examination? Will I get writer's cramp?' You will have done enough study in anatomy that hands do not fall off because of writing for a few hours, even if it may feel like it! All physical skills need practice, and if an exercise is performed without months of preparation, muscles ache. So train your body by practising writing out the answers in longhand, to build up your hand muscles. If you are taking a viva voce examination, ask a friend to question you with some sample questions.

Show You Know

We have already talked about a number of ways in which you might show other people what you know, but the important thing is that you convince yourself. There is no point reading information and then saying to yourself 'Yes, I know that' and moving on.

You know something if you can recall it whenever you want – you cannot be sure that you know it just because you can recall it 30 seconds after you looked at it on a piece of paper in your hand. Test yourself by trying to recall important pieces of information when you are a long way from the 'crutch' of your revision aids. *Learn* and *relearn* – go back over lists several times to be sure that you know. You must demonstrate to yourself that you do know the information. How do you do this? It is up to you. Recite things out loud, write down lists and stop yourself from eating that next biscuit until you can recall everything that you need to – whatever it takes. Talk to your family and friends: teach them your anatomy knowledge – get them as fired up and enthused about anatomy and physiology as you are.

Get to recognise your self-destruction tactics and stop them in their tracks. Procrastination, self-doubt and apathy are your own choice, so choose not to do them!

Above all, make it FUN! There is no rule to say that revision has to be stressful, miserable or boring. Your revision experience is your choice – it's up to you, so aim to enjoy it!

Examination Toolkit

1. Sort out *your equipment* the day before. Put pens, coloured pens, pencils, ruler and eraser into a transparent pencil case.

2. If you need to *bring identification* with you, pack it the day before and keep it with your pencil case. This all helps you to remain calm and organised if you have already packed everything you need.

3. If you are allowed to bring sweets or food into the examination hall, take note! If you must eat sweets during the exam, make sure you pick quiet ones. It sounds a daft thing to say, but when you are deep in 'examination-mode' and you are scribbling away like a mad thing, there is nothing as off-putting as a fellow student unwrapping a sweet with rattly, crinkly paper and then chomping away, slurping and crunching! *Unwrap your sweets* before the exam starts, and choose sweets that melt in the mouth.

4. Do *have breakfast* on examination day. Feed your brain! Saying 'I don't eat breakfast' is like saying 'I don't put fuel in my car.'

5. Don't get stuck into files at the last minute – *use brief revision aids* instead, if you want to do last-minute revision

6. *Set the alarm clock a bit earlier* than you need. Set two alarm clocks if you struggle with early waking.

7. Get to the examination in *plenty of time.*

8. *Keep calm and breathe!* Avoid panic by keeping your breathing slow and deep. If panic begins to rise, take a breath. Put one hand on your sternum, and the other over the umbilicus. Concentrate on inhaling through the nose and exhaling through the mouth, whilst monitoring the movement of your hands. Breathe so that both hands move equally. A wonderful panic averter. Concentrate on feeling the floor with your feet. What can you feel? Can you feel your shoes? Socks? Give your toes a wiggle. By concentrating on something else other than the panic, the terror will disappear, and you will feel grounded and more able to continue the exam.

9. *Drink water.* Hydration is key for good brain function; your textbooks have told you that when you were revising. Keep that brain hydrated with sips of water so thought processes are clear.

10. Always *read through the examination paper* before answering a single question.

Why do I have to read through the exam paper before I start writing? Shouldn't we just crack on with answering the paper?

Read on for an example in just why we advise that you read all the exam paper first:

Please read all of the instructions before doing anything, you are allowed ten minutes to complete this task.

1. Find a pen and paper.
2. Write your name at the top of the paper.
3. Write the numbers 1 to 5, one per line.
4. Draw five small circles beside #1.
5. Put an 'X' in the second and fourth circles next to #1.
6. Write the word 'encyclopaedia' beside #3.
7. On the back of the paper, multiply 7 × 9.
8. Put an 'X' in the lower right-hand corner of the paper.
9. Draw a circle around the 'X' you just made.
10. Underline your name.
11. Say your name out loud.
12. Draw a circle around #4.
13. Count the number of words in this sentence and write the answer beside #2 on your paper.
14. Put a square around #1 and #5.
15. Punch three small holes anywhere in the paper.
16. Write your first name beside #4.
17. Write today's date beside #5 on your paper.
18. Circle every letter 'E' you have written.
19. Stand up and say 'I have finished first' if you were first, else say 'I have finished' out loud, then sit down.
20. Now that you have read all of the instructions, skip all of them except the first two. If you have followed the instructions correctly, you should only have your name on the paper!

11. *Avoid waffle and vagueness.* Examination questions are designed to see if the student can work out what the question is asking. It is very upsetting for an examiner to read a beautiful piece of prose containing perfectly recalled knowledge, only for it not to answer the question that was actually asked. Also, avoid using terms such as 'etc.' and 'and so on'. Be specific! The examiner is not a mind reader so won't give you marks for 'etc.' and or 'and so on'. Give bullet bullet-pointed lists instead. Examiners are also fine-tuned to detect when a student has reworded the same sentence three different ways in order to gain marks. Although it is a good approach to assume the examiner knows nothing, they do really, and can sniff out repetition!

12. *Clock-watch.* Plan the timing need so you are clear how many minutes you should devote to each question, and keep a watch to hand so you can keep glancing at the time. Work out how many questions you should have answered by halfway through, so you can move on to the next question if time is ticking too quickly for you. Give yourself ten minutes at the end of the examination to read through the questions and to check your answers. If you think of a few extra things to say, or you are running out of time, use bullet points and lists to scribble the answer down.

13. *The very beginning is not always the best place to start.* Examinations may be structured with some shorter-answer questions in the first section, followed by higher-marked, longer, essay-style questions. It is easier to get the bulk of the marks in the first few minutes of answering the essay questions, so it is my advice to do the high-marked questions first, and the one- or two-marked questions last. That way, if you run out of time, you will have managed to accrue some good scores. You will have read through the examination questions first, so will be aware of the allocation of marks.

14. *Answer all the questions that you need to answer.* You won't be marked down for putting a wrong answer, and you will only get marks awarded if you give the correct information, so when in doubt, write it down. Read the questions carefully, marking the key parts of the question to remind yourself to address all the points. As you answer a question, cross its number out on the question paper. You will then easily see if a question has been missed out. Do not be tempted to score through your rough notes. If you run out of time, your bullet points and rough scribbles may be included as your answer if they are written in the answer script or in the space provided for your answers, so long as they are not crossed out. Anything written or drawn that has been crossed out is usually ignored by the examiners.

15. Use *diagrams* to explain yourself. The examiners are not expecting fine art, so a rough sketch is usually fine. Tables can be a useful way of clearly laying out answers, particularly for 'compare-and-contrast'-style questions.

16. Keep a *positive frame of mind.* If you watch athletes at the start of a race, they are telling themselves that they are going to succeed. You have done all the work and deserve to succeed. Believe that you can too!

And finally ... **GOOD LUCK**!

Avoid Common Examination Mistakes

(Adapted from The Tutor Pages (2013).)

1. *Disregarding the marking scheme*

   Only writing two sentences for a five-mark question is not going to get the necessary points. If the marking schemes says two marks, it means that there are two or four pieces of information that are required, and so on. Ten-mark questions require ten different pieces of information – not the same piece of information reworded ten times!

2. *Writing too much for the short answers*

   Short-answer questions need simple sentences or bullet points that get across the key points. Check the number of marks that are allocated, and answer accordingly. Don't waste half an hour scribbling five pages of detail, when there are only two marks allocated to the question.

3. *Confusion over the words 'describe' and 'explain'*

   This is one of the most common examination technique errors. Problems include:

   - Describing something inaccurately or with insufficient detail
   - Omitting units from quantitative descriptions
   - Explaining something instead of describing it
   - Describing something instead of explaining it
   - Explanations which are too short and lack detail.

4. *Failure to use comparative terminology*

   Biological systems are complex and adapt to changes. Understanding that these changes must be described using comparative terms such as faster, slower, more, less, fewer, taller, greater, shorter, brighter, and so on, is key to writing good answers. A statement such as 'The atrioventricular valve opens because pressure in the atrium is high,' will not gain full marks. The use of a comparative term is crucial: 'The atrioventricular valve opens because pressure in the atrium is higher than pressure in the ventricle.'

5. *Confusion over 'structure', 'properties' and 'function'*

   Biological structures are exquisitely adapted to carry out their functions. Their structure determines their properties, which in turn allow them to function efficiently. Though closely linked, these three terms have distinct meanings and cannot be used interchangeably. When asked a question such as 'Describe the structure of a collagen molecule,' candidates will often describe its properties and function, e.g. collagen is 'insoluble, tough and gives strength to artery walls', rather than its structure.

6. *The use of vague and imprecise terminology*

   Science is full of specialist and technical language which many students find difficult to master, and you will need to be able to 'speak anatomy,' as was seen in Chapter 2, earlier. Together with the fear of plagiarism, and the instruction to 'put ideas into your own words', it is little wonder that students write imprecise and vague answers. There are many examples of this, but perhaps the use of the word 'amount' is the most common. When asked to state factors which must be kept constant in an investigation, examination candidates will often write vague statements such as 'the amount of sodium chloride solution'. In physiology this is meaningless – the term 'concentration' or 'volume' should be used instead. Other specialist terms to use instead of 'amount' could be 'mass' or 'number'.

7. *Misspelling*

Although anatomy and physiology are not a spelling test, there are a few words which can be easily confused, and where misspelling a word means that the term you use is in a completely different part of the body than where you meant. This is vitally important when it comes to medical matters, so best learn the correct spelling of the anatomical parts now.

Ileum is in the intestines, and ilium is a pelvic bone. Don't mix them up!

Trapezius is a muscle, and trapezoid is a ligament in the shoulder as well as a carpal bone in the wrist, and trapezium is also a carpal bone.

Perineum is the area between the genitals and the anus, whereas the peritoneum is the membrane covering your intestines and lining the abdomen.

Perineal is the area around the perineum, which is between the genitals and the anus, whereas peroneal refers to the area around the peroneus muscle, which is also known as the fibularis muscle, on the outside of the lower leg. Please don't confuse the two.

Trachea is the tube between the back of the throat and the lungs, whereas trochlear is the name of the fourth cranial nerve, which innervates the eye. And just to further confuse, the trochlea is a term in anatomy which means a grooved structure, similar to the wheel of a pulley. Trochleae are found at the humerus, femur and talus.

Final Thoughts

It can be seen that there are revision and preparation methods that act as scaffolding for your brain, supporting and networking information to make it memorable, and examination techniques to guide you through any potential minefields of confusion and to help it make sense. Use your resources wisely and appropriately to make your 'learnscape' your own.

Preparation for examinations is a continuous process which requires time and concentration. This chapter has provided different study and revision techniques to aid the revision process. The challenge is to use these techniques to develop examination skills and demonstrate knowledge under different assessment conditions. The authors wish you the very best of luck in your examinations and assessments, and hope that your success is only the start of your lifelong passion for learning anatomy and physiology.

References

The Tutor Pages (2013) The Importance of Exam Technique. Available online at: http://www.thetutorpages.com/tutor-article/a-level-biology/the-importance-of-exam-technique/6318. (Accessed 15 June 2015.)

University of Bath (2015a). Avoiding Procrastination. Available online at: www.bath.ac.uk/students/support/academic/procrastination/index.html. (Accessed 13 March 2015.)

University of Bath (2015b). *Time Management.* Available online at: www.bath.ac.uk/students/support/academic/time-management/index.html. (Accessed 13 March 2015.)

7 Test Anatomy

He that knows not,
and knows not that he knows not,
is a fool.
Shun him.
He that knows not,
and knows that he knows not,
is a pupil.
Teach him.
He that knows,
and knows not that he knows,
is asleep.
Wake him.
He that knows,
and knows that he knows,
is a teacher.
Follow him.

Arabic proverb

You may have turned to this chapter because, as a student, you are curious to see how a teacher prepares you for your exam. Well done for getting this far! Read further to gain an insight into the preparatory steps taken by teachers to assist with student revision, as well as into how teachers and assessors develop the questions for tests and examinations.

If you are a teacher of anatomy and physiology, use this chapter to help you prepare your students for assessment and examination, and to help you prepare good and fair examinations. But do take some time to read Chapter 6 – Learn Anatomy, to see things from a student perspective too. Walk a mile in their shoes, so you can understand their concerns, frustrations and fears when it comes to the all-important examination process. In this chapter we concentrate predominantly on written assessments, but acknowledge that anatomy and physiology can be examined in multiple formats. We therefore offer a few suggestions regarding spoken examinations as well.

Helping Students to Know what They Need to Know

The role of a good teacher is not necessarily to teach the students what they need to know, it's to teach the students to work out WHAT they need to know, and to recognise that they do *not* know it. It can also be described as being in conscious incompetence. Total unawareness of what is actually required to be learned is a dangerous place to be when studying – ignorance is *not* bliss. Learners may face some difficulties, barriers and conceptual bottleneck situations in learning anatomy and physiology. To help identify these problems, read Chapter 5 – Coming Unstuck.

How Do Students Benchmark what They Need to Know?

In the first instance, look for a list of what a student needs to know. This will be found in the *intended learning outcomes* (ILO), which should always be accessible from the course provider, and are a key component of course planning and assessments for students and the teachers.

A good examination paper is based on detailed ILO. Teachers use these to set the content of the lessons, the level of depth of knowledge and the assessment questions. Students will use ILO to check that they have been taught everything they need to know, and should use them as a tick list in their revision strategies, so it is important that these are set as foundation stones to building the course and the examination. This process is known as *constructive alignment* (John Biggs, 1999). The ILO set a minimum standard for what is acceptable at the end of the course, i.e. the threshold level for a pass, and should clearly express the expectations of the students' understanding of the subject.

For all teachers who are busy preparing the anatomy and physiology courses, think of the learning outcomes as a series of steps (from *Bloom's Taxonomy of Educational Objectives* (Atherton, 2005)). For all students, use these steps to monitor your learning as your course progresses, perhaps making up your own mock examination questions.

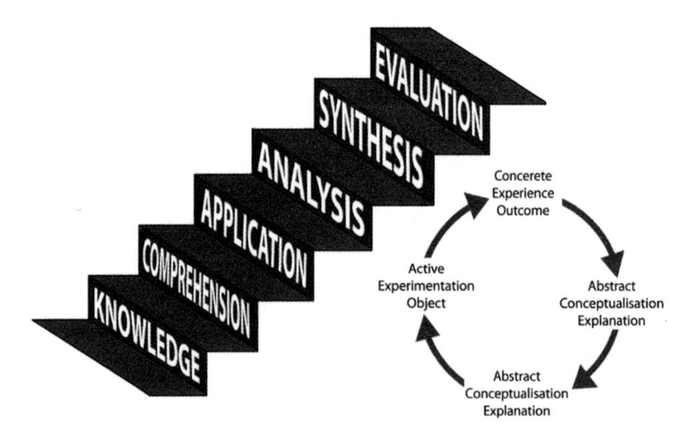

Figure 7.1: Think of learning outcomes as a series of steps. Adapted from Atherton (2005).

Step 1 – Knowledge

Applying the verb 'to know' to a learning outcome is far too vague. It is much better to describe what a student can do with the information to demonstrate their knowledge (Gronlund, 1998).

So, instead of 'know', use words such as:

- Recognise
- Define
- Name
- List
- Describe
- Identify
- Match
- Select
- Recall

These types of learning objectives are perfect for thinking up short-answer questions such as defining anatomical terms, or listing the functions of an organ.

Examples:

These are three versions of a question to assess the knowledge of the parts of the femur, which approach the assessment in different ways, thus requiring a different way of processing the information.

***List* the anatomical landmarks of the femur.**

Does the student have the ability to recall the names of the landmarks of the femur?

This does not necessarily test the application of that knowledge, just simply can they regurgitate a list?

***Identify* the anatomical landmarks of the femur.**

Can the student actually recognise these landmarks and label a diagram?

This would demonstrate that the student does indeed know exactly where the landmarks are located, rather than just knowing their names.

***Describe* the anatomical landmarks of the femur.**

Can the student describe the landmarks?

This requires much more detailed knowledge than simply being able to point to a location and name it. It would also require a higher level of anatomical language.

Step 2 – Comprehension

Comprehension of anatomy makes the information useful, but just asking if a student 'understands' a topic is not precise enough. After all, how can students check their own understanding when the depth of understanding needed to pass is too vague? Instead use terms such as:

- Recognise
- Report
- Review
- Explain
- Summarise
- Distinguish between
- Give examples
- Illustrate
- Classify

These verbs can be used in examination questions that require a short paragraph in answer.

Examples:

Give examples of the endocrine and exocrine glands.

Is the student familiar with the terms endocrine and exocrine?

If so, they would be able to give a list of each, correctly subdividing them into the appropriate category.

***Distinguish* between the endocrine and exocrine glands.**

Can the student remember the definitions of the endocrine and exocrine glands?

The understanding of the functions of the two systems and how they differ from each other is required here.

***Classify* the organs of the endocrine system based on the type of hormone being produced.**

Can the student remember the names and functions of the hormones produced and their biochemical nature?

A much deeper understanding of the endocrine system and its hormones would be required to classify the organs. Biochemistry of the hormones and their functions and actions is a very high-level knowledge compared with the other examples given above.

Step 3 – Application

Combining knowledge and understanding brings a higher level of learning where application of the information can be used relevantly. This can be assessed by using words and verbs such as:

- Interpret
- Solve
- Use
- Employ
- Arrange in order
- Show

These can form examination questions based on case histories and medical conditions, where the symptoms are due to dysfunctions in the anatomy and physiology.

Examples:

A 45-year-old lady has swollen joints of the hands and toes and early morning joint stiffness which remains until late into the morning. She has been very tired recently and has complained of a low-grade fever. Which clinical condition needs to be investigated in order to solve the cause of her symptoms?

Is the student able to reflect on the symptoms being listed and able to apply them to a clinical condition?

The combination of knowledge of rheumatoid arthritis and the understanding of its symptoms is what is being assessed here.

Show **how the symptoms of inflammation are caused.**

Can the student explain the application of the information that has been learned sufficiently to put the explanations into his or her own words?

This answer requires a list of the symptoms of inflammation and an explanation of why these processes cause inflammation, which is altogether more detailed and in depth than simply recognising or listing the symptoms

Use **the following terms to explain the cause of rheumatoid arthritis symptoms:**

- **Auto-antibodies**
- **Synovial membrane**
- **Cytokines**
- **Rheumatoid factor**
- **T-cell activation.**

Can the student understand the cause of rheumatoid arthritis sufficiently to correctly use these terms?

Although, on first glance, you may think that this is an easy question as some words have been given to help you answer it, those terms that have been provided impart the level of knowledge that is expected. If the terms had been 'auto-immune', 'joints', 'swelling' and 'hot', the required depth of knowledge would have been far lower.

Step 4 – Analysis

Deeper analysis of the relationships between the body systems requires learning outcomes, using verbs such as:

- Analyse
- Compare
- Contrast
- Criticise
- Appraise
- Experiment
- Differentiate
- Order
- Subdivide

'Compare and contrast' structures or organs is a useful way of seeing if a student has understood and applied the information sufficiently to reach an informed analysis.

Examples:

Differentiate between the structures of the male and female pelvis.

Does the student fully understand the need for the structure and function of the pelvis to differ according to gender?

If so, they would be able to apply the knowledge of the reproductive system to the knowledge of the anatomy of the pelvis.

***Compare and contrast* the structures and functions of bones of the male and female pelvis.**

Is the student able to see the similarities and differences of the male and female pelvises and able to apply that information?

Knowing that pelvic structure differs between the genders is insufficient. This question requires the student to understand the functions of each of the bones, and to apply knowledge of the reproductive system to this.

***Analyse* the requirements of the female pelvis.**

Is the student able to work out the body systems that are located within the pelvis, and to identify the need for structure to be defined by function?

In a similar way to the last question, an analytical approach is required to reach the depth of understanding needed to answer this.

Step 5 – Synthesis

The putting together of all the information can be assessed by the ability to construct an argument and integrate knowledge. Use verbs such as:

- Develop
- Organise
- Construct
- Create
- Design
- Compile
- Plan
- Write

Examples:

Heart rate during a period of exercise is linked to the intensity of exercise. Create a flowchart showing the interactions between the respiratory, circulatory and nervous systems during exercise.

Can the students link the functions of the body systems to each other during exercise?

This task requires cross-referencing systems, and applying their functions to an event.

Design **a controlled experiment to determine the relationship between intensity of exercise and heart rate.**

Can students extend their understanding of the interaction between the body systems beyond listing and cross-referencing?

Designing an experiment to prove relationships between cause and effect requires a very high level of understanding and application.

Construct **a graph of the results you expect for both the control and the experimental groups for the controlled experiment described above.**

Is the student able to predict the outcome of their experiment?

Predicting the expected results would show an even higher level of learning beyond understanding cause and effect.

Step 6 – Evaluation

This is thought to be the highest level of learning, demonstrating that the students can have their own opinions based on all the evidence learned. Ideally, this requires essay-style answers, and can be assessed by using verbs such as:

- Assess
- Estimate
- Theorise
- Argue
- Appraise
- Judge
- Reflect
- Verify
- Discriminate

Examples:

Theorise why proponents of barefoot running argue that running without shoes is a more natural form of exercise.

Can students defend their arguments and justify what they believe to be true within an argument?

The questions ask students to look at one side of the discussion and apply all their knowledge of structure and function of the foot to create theories to support a viewpoint.

Write an informed argument **to show that running without shoes is a more natural form of exercise than running in shoes because of the anatomy of the foot.**

Is the student able to see both sides of the argument?

Although seemingly similar to the first question, an informed argument requires both sides of the argument to be discussed, compared, contrasted and presented within the answer.

Verify **the type of footwear which has the most natural effect on the foot and its anatomy.**

Is the student able to apply the effect of an artificial object such as footwear to their knowledge of foot structure and function?

In order to determine the truth of a postulation, all relevant information must be considered, at the highest level of learning, to produce a cogent, cohesively argued conclusion.

The best anatomy and physiology course is one in which the students know what they need to learn, and, more importantly, they know what they don't know. In other words, they know what they need to still learn. Learning should be tracked via self-assessment and completion of a training and learning log.

This can be based on the intended learning outcomes (ILO) as a tick list or checklist. Give the responsibility of completion of the training log to the students – it encourages them to keep track of their own learning.

If you are a student, get that checklist now and create your own detailed checklist and training log based on suggestions in this book.

Posing the Questions

As mentioned earlier, the ILO provide the minimum level of competence required, so the formation of these is crucial in working out minimum standards, and the all-important pass mark. Some examinations require only 40% to class as a bare pass, whereas others require that the student gets more right than wrong to achieve a pass grade.

In order to be fair, the question bank and examination paper should have a spread of questions at varying degrees of difficulty:

- Simple and straightforward to introduce the examination and to allay any nerves, making up about 15% of the examination
- Standard, middle degree of difficulty, that will probably make up about 60–70% of the questions
- Harder questions with more depth of knowledge needed, to extend the very able student, making up about 15% of the total questions.

The questions should be clear and unambiguous. Nothing is worse for panicked students than to read an examination question and not even know what that question means or asks for. Ideally, if the question is a lengthy one, and has several parts to it, the question should be broken down into subsections, which makes it much easier for the student to realise the requirements of the examiner.

Poorly Written Questions

These are some common practices that result in poorly written anatomy and physiology questions (adapted from Boonshoft School of Medicine (2015); portions included with permission from the National Board of Medical Examiners:

Unnecessarily complex questions and/or directions – Unnecessary complexity may be due to the amount of information included in the item, to poor grammar, or both.

Not specifying the basis for sequencing/ordering – The basis for ordering things should be specified in the assessment item. For example, putting muscles in a certain order can be done in several ways: by distance from the core, by size or by alphabetical order.

Not indicating the nature of the desired description – Most things can be described in at least two ways: by their physical features or by their functions. Test items should indicate both what is to be described and what is to be included in the description. An example of this could be describing the central nervous system: listing its functions gives a completely different answer from listing its structure and components. Be specific in the questioning – 'Describe the functions of the central nervous system,' or 'Describe the structure of the central nervous system and its components.'

Using absolutes – The use of absolutes (e.g. always, never, only, no) should be avoided. Not only are absolute statements usually incorrect, but also some students will know an exception to the keyed answer and get confused or panicked.

Using implausible alternatives – When multiple-choice items contain alternatives that are obviously incorrect, students have a greater chance of selecting the correct choice.

Using equal numbers of items for matching items – When the number of items to be paired in matching items is the same it increases the chance of students guessing correctly on items not previously learned.

Including grammatical clues – When asking a question that includes filling in the gaps in a sentence, articles such as 'a' and 'an', plural word forms, and gender forms may provide clues to correctly answering without having learned the content. Reword the sentence to remove these clues.

Having no reference to alternatives – The students should be able to read the question and answer it without reference to the alternatives – and their answer should be among the alternatives when they get there. In other words, the stem asks a question that has a definite answer.

Not emphasising key words – Anytime you have a key word that would change the answer if the student misses it, **boldface** it, CAPITALISE it, <u>underline</u> it, *use italics* or ***<u>all of the PREVIOUS</u>***.

Not specifying what best answer is – When best-answer items are used, the qualifier in the stem needs to be emphasised, and also you must specify in what way one answer is the 'best'.

Negatively worded questions – Negatively worded stems should be avoided. They are literally a no-no! Use positive language. Instead of asking 'Which endocrine cells in the pancreas do not produce insulin?' ask 'Which endocrine cells in the pancreas produce glucagon?'

Lengthy information in alternatives – Include as much information in the stem and as little in the options as possible. For example, if the point of an item is to associate a term with its definition, the preferred format would be to present the definition in the stem and several terms as options rather than to present the term in the stem and several definitions as options.

Using 'all of the above' – In multiple-choice examinations, avoid the use of questions that require the answer 'all of the above.' Recognition of one wrong option eliminates 'all of the above' and recognition of two right options identifies it as the answer, even if the other options are completely unknown to the student. 'All of the above' can also be unfair, because it means if answer (A) is correct, the students anticipate (A), look down, see (A), choose it, then move on – getting it wrong, even though they knew (A) was the correct answer. This means students can get an answer right but overall their answer is wrong for that question. It's just not fair to do this.

Using trickery – The use of trickery must be avoided. There is no point in finding out if you can trick the students – again, it's not fair to try to trick an already nervous student. You are not assessing students' ability to be tricked.

Use sensible marking allocations – The allocation of marks should be fair and unambiguous, and should make it clear how much information is required in the answer. For instance, a ten-mark question about the digestive system will require at least ten different facts, or five facts with an example of each, and a two-mark question requires two facts, or four in less detail, or two facts and an example of each.

The best questions allow differentiation. This means that a question is asked that can be answered in a very basic way to get minimum marks, with more detail to gain a pass mark, and with extension opportunities so that the high fliers can show off their knowledge.

Example:

Describe in detail the process that occurs at a neuron to allow the action potential to occur

A simple answer to this would list the processes in very simple terms, and would receive fewer marks than an answer that went into detail about the physical structure of the neuron and how this is designed to do the task, with full details of sodium and potassium ions, resting potential, action potential and refractory period.

Tables

Tables are a great way of organising information for revision and a super way of asking anatomy and physiology questions. They are easy to mark, so a great favourite for the examiner!

Example:

Complete the following table: state the endocrine organ that produces the listed hormone

Hormone	Organ
Insulin	
Oestrogen	
Testosterone	
Adrenaline	
Antidiuretic hormone	
Growth hormone	

Diagrams

Anatomy requires a visual knowledge of the body at a gross-anatomy level (normal size) and at a microscopic level. Drawing or labelling diagrams really does test that knowledge.

Examples:

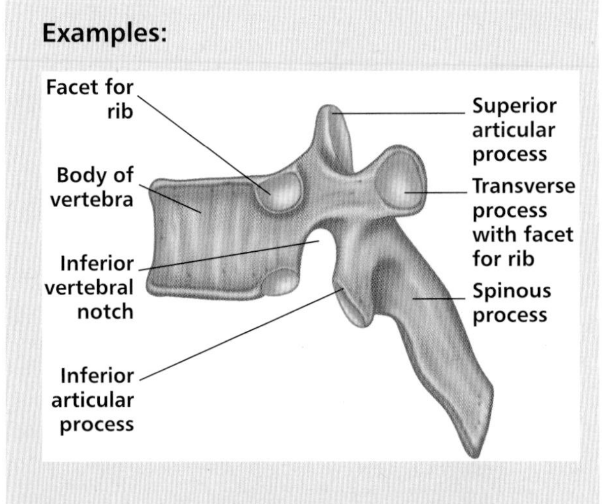

Figure 7.2: Landmarks of a thoracic vertebra.

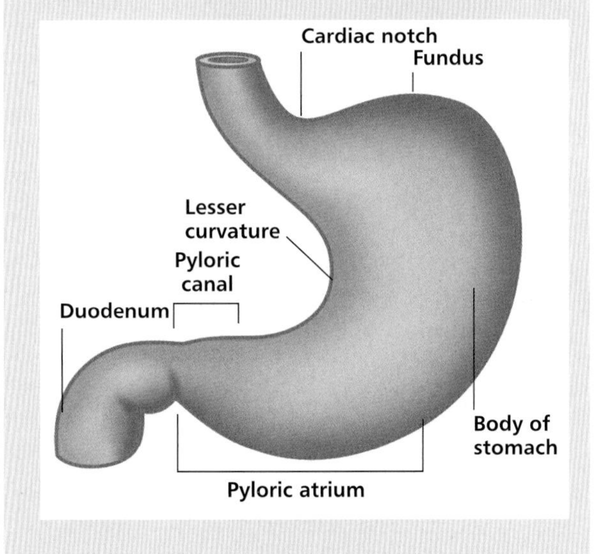

Figure 7.3: Cross section of stomach.

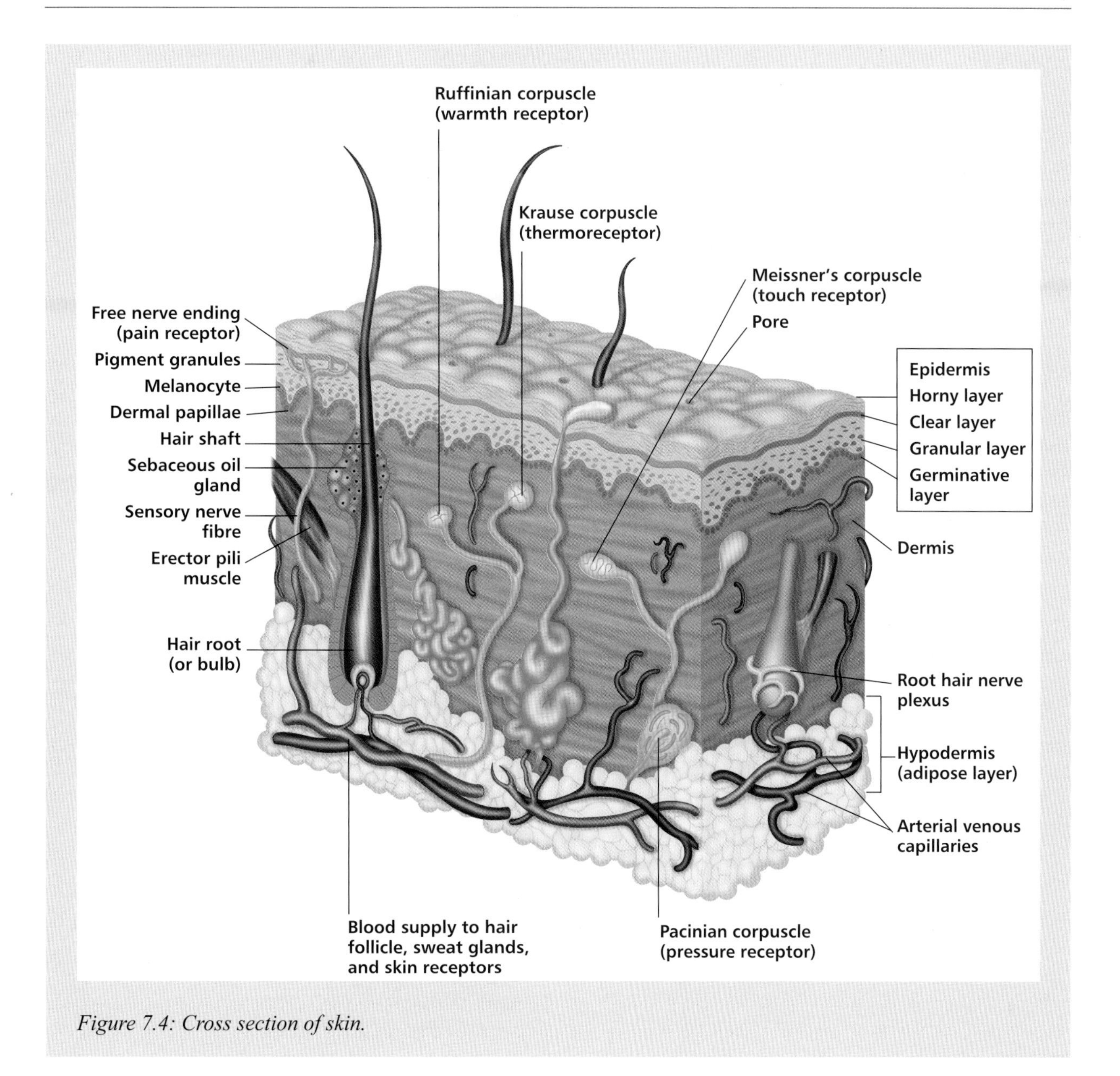

Figure 7.4: Cross section of skin.

Essays

These should be set according to the ILO. Verbs and other words used to frame the question should show a spread of the learning levels according to Bloom's *Taxonomy of Educational Objectives* (Atherton, 2005), so that there are some questions that test each different level of learning. This was described in detail earlier in this chapter.

Short-Answer Questions

Some lower-marked questions should also be used, mainly to test the lower levels of learning and recall. These are useful as compulsory questions, rather than ones that are chosen. Short-answer questions can test some dusty corners of the syllabus that get ignored too.

Complete the Sentence

These can be popular as homework questions, but may not be an effective assessment technique as they do not really test knowledge at a significant level. More effective are questions which ask the student to write their own paragraph as an essay or short answer.

Multiple-Choice Questions

A multiple-choice question (MCQ) examination is seen as simpler to pass by many people, but a properly constructed MCQ should be as testing as an essay-based examination. The format is useful for large numbers of students where marking lengthy written papers is not a cost-effective use of time. It also lends itself to peer-marking for mid-course assessments. It does, however, require the most work in terms of creating a question bank.

Pros of MCQ	**Cons of MCQ**
Scoring is easy, objective and reliable	Constructing good questions is time-consuming
Learning outcomes, from simple to complex, are measurable	Does not allow assessment of problem solving as well as essay-style questions do
Good for large numbers of students	Scores can be influenced by reading ability
Suitable for peer-marking	Difficulty to test the higher levels of cognitive thinking
Covers a lot of material efficiently (about one mark per minute)	May encourage guessing
Questions can be written so that students must discriminate among options that vary in degree of correctness	Sometimes more than one answer is correct

The question bank should be planned, and a blueprint of the test should be produced which ensures the desired coverage of topics and that the level of objective is met:

- The number of questions required for the examination
- ILO being assessed
- Skills and learning levels being assessed
- Relative weight of marks given to each section.

Closed-Book Examination

The traditional examination assesses the ability to memorise information as well as apply it. The drawback to this is that students will focus on memorising the information in order to take and pass an examination, rather than exercising the application of the information, which would make it more memorable in the long run.

Another drawback is that it relies on getting memorised information down as fast as possible, so those students who have worked diligently throughout the year, yet do not have such a fast processing speed, struggle to write as much as they should. An examination should test knowledge and understanding, rather than memory and speed of recall and writing.

The advantages in a well-planned closed-book examination are that the questions test memorised understanding, and that memorised understanding is retained forever if practised and revisited regularly.

Open-Book Examination

Taking an assessment with a textbook in hand is an unfamiliar concept to many. Already knowing the title of the essay that has to be written? Sounds like a walk in the park? Think again. An open-book examination ensures that the subject matter is well researched and revised before the assessment. And if the textbook is allowed in the examination as a reference, it promotes cross-referencing. Some open-book examinations involve writing an essay the title of which has been made available a set time before the assessment.

An open-book test allows for and expects expansion beyond the information already delivered. The skills learned from open-book assessments are likely to help students in their learning and give them key skills:

- Organising of notes and locating of topics within a textbook
- Working within a time limit
- Learning that preparation aids success
- Engaging debate and active learning.

It is sometimes thought that an open-book test is actually more representative of the workplace, where a textbook could be fetched down from the shelf rather than all the information having to be committed to memory. Others argue that memorising the information is more important and that the provision of resources in an examination simply teaches how to locate information rather than how to apply it.

An approach to this type of assessment would be to use the index of the book to identify the pages that give the most information about the subject, and to mark those pages with sticky tabs to make them easier to find. Label the sticky tabs with the topic in question so they can be located easily.

Far too many students leave the preparation of the answer to the last moment, so do not use the study time appropriately. Research and plan the essay as if it were a homework question. Then 'chunk down' the essay to form memorable paragraph titles which will prompt the full answer when in the examination.

Spot Tests

Spot tests are assessments which often use a 'prop', such as an anatomical model of a body part, or a diagram/photograph of a cadaver prosection. Like the closed-book examination, they do rely on memory. The test would involve looking at a structure, photograph, diagram or model and identifying landmarks and specific anatomical features.

As well as using this as a written examination, it works well for small groups as regular assessment. It can be peer-assessed – the rest of the group discuss the question and decide amongst themselves which is the correct answer.

Example:

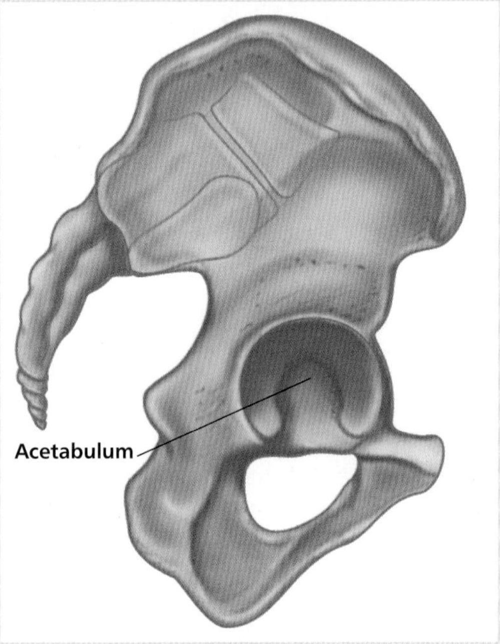

Acetabulum

Figure 7.5: What is the name of this structure in the pelvis?

Viva Voce (Oral Examinations)

Oral examinations are often used to assess anatomy and physiology in higher education environments. Some awarding bodies may also consider using oral examinations for students who have a specific learning difficulty such as dyslexia or dyspraxia, where the ability to manage to read and write would be under test, rather than the knowledge of anatomy and physiology. They may also be used by examining bodies as a method of final assessment in the form of an interview. Each examining body will have their own format and regulations, and you are advised to consult your examining body for advice in format and structure, including lists of sample questions.

An oral examination of anatomy and physiology involves an examiner verbally asking the questions and the student saying the answer out loud. Administration of the examination depends on the rules of the awarding bodies, but usually the entire examination is recorded, either audio or video, and a scribe writes down the answers given by the student.

Anatomical models and skeletons are useful to have available for the student to hold, thus allowing them to demonstrate the location of certain landmarks and structures.

Feedback and Feedforward

Feedback provides information to learners to identify gaps or misconceptions in their knowledge, enabling them to evaluate their progress and to give an indication of where they are in relation to their learning goals and objectives. Feedback is a vital component of effective learning. It typically looks at the current performance of the learner and can be given by teachers, supervisors, mentors and peers. It can also be given by computer and as a result of self-assessment. It relies on the learner reading the comments given as feedback, and acting on them in subsequent tasks. David Nicol and Debra Macfarlane-Dick identified seven principles of good feedback practice (Nichol and Macfarlane-Dick, 2006):

1. Clarify what is good performance (goals, criteria, expected standards).
2. Facilitate self-evaluation and reflection.
3. Deliver high-quality feedback information.
4. Encourage peer and teacher dialogue.
5. Encourage positive motivation and self-esteem.
6. Provide opportunities to close the gap between current and desired performance.
7. Use feedback to improve teaching.

In order for feedback to be effective, it must be timely, allowing time for the learner to integrate the comments and advice and put feedback to good use.

Implementation of feedforward assignments can be a useful method of guiding learners to reach the desired learning goals. Such feedforward tasks could be in the form of submitting an essay plan of a particular topic. Feedback is given on the plan, and learners immediately incorporate the feedback into their preparation for the essay, giving scaffolding for their understanding and enabling them to consolidate their learning, so that they finally achieve a better-prepared assignment.

Final Thoughts

Assessment of learning is an essential part of documenting the progress of both the learner and the teacher. Far more important, though, is thorough, detailed and appropriate assessment conducted in a fair and just manner with feedback and feedforward, allowing learning and knowledge to flourish. Just as a tiny cell needs the appropriate nutrients in its environment to reproduce and grow, our assessment techniques need to demonstrate both growth and application of knowledge. This will enable the teacher to monitor how, and whether, learning outcomes are being satisfied.

References

Atherton, J.S. (2005). *Learning and Teaching: Bloom's Taxonomy*. Available online at: www.learningandteaching.info/learning/bloomtax.htm. (Accessed 11 June 2015.)

Boonshoft School of Medicine (2015). *Academic Affairs: Test Items*. Available online at: https://medicine.wright.edu/academic-affairs/faculty-development/test-items. © National Board of Medical Examiners. (Accessed 11 June 2015.)

Gronlund, N.E. (1998). *Assessment of Student Achievement*, 6th edition, Allyn Bacon.

Nichol, D.J. and Macfarlane-Dick, D. (2006). Formative assessment and self-regulated learning: A model and seven principles of good feedback practice. *Studies in Higher Education* **31(2)**: 199–218.

8 Own Anatomy

The Anatomical Toolkit

This chapter considers how the reader develops ownership of understanding and applying anatomical knowledge through the use of a conceptual toolkit. The chapter is divided into two sections. The first discusses the development of the anatomical toolkit, whilst the second explores reflective practice. The content of the chapter enables the reader to better develop connections between ideas and enhance learning and teaching strategies for anatomy and physiology.

Developing the Anatomical Toolkit

Figure 8.1: The anatomical toolkit.

The anatomical toolkit is a metaphor for building anatomical knowledge, i.e. the structural components of anatomy are like a toolkit carefully placed and developed. The toolkit concept has been used in a multitude of ways throughout this text. Chapter 9 – Resource Anatomy explains the usefulness of the tools in the toolkit, whereas this chapter examines how to use the toolkit to build knowledge. Both this chapter and Chapter 9 have embedded the term toolkit to illustrate the intricate balance between process and product.

The Concept of a Toolkit

Have you ever attempted a repair job without the right tools? This can be emotionally frustrating, awkward and lead to further expense. But having a toolkit is one thing, knowing how to use the tools effectively is something different. A toolkit is more than a collective space for storing tools – it provides a central hub for locating tools and accessory parts. The toolkit alone is insufficient without the knowledge and appreciation of how the tools work individually and collectively. With this in mind, let us examine the structure of an anatomical toolkit from its component parts to understanding the whole, like a giant puzzle whose pieces we will attempt to segment together.

I am often amazed watching a skilled tradesman working meticulously with a range of tools, each one used to perfection, each one carefully placed in the toolkit, each one checked for operational defects. Tools are wonderful things to have and develop, whether they are used for construction, cooking, health, learning or application, the range of tools and proverbial toolkits is vast. So what exactly is an anatomical toolkit and what purpose does or should it play in building anatomical knowledge?

The chapters of this book have considered different learning and teaching ideas, ideologies, activities, metaphors and tasks; however, the common thread that weaves all these ideas together is how each idea works alone and in concert with other ideas. The toolkit provides the foundation for recognising and identifying each tool, understanding how each tool is shaped, and knowing the purpose and operation of the tools. In anatomy, this is similar to building the human body. Imagine, if you can, that one of the first tasks you are asked in an anatomy class is to build a model of the human body. You may decide to build the skeleton to depict the anatomical framework, onto which you could layer different tissue and visceral structures. If we use the skeleton as a starting point, the first concern is selecting the material. Will the skeleton be built out of paper, plastic, wood, iron, wire or cardboard? The selection of material will determine which tools are used in the construction of the model. If wood is used, you may select a saw, drill and drill bits, screws, screwdriver, nails, hammer, glue, and hooks. The design of the skeleton will further determine the specific use of tools and accessory hardware. If your toolkit contained only a hammer and a few nails, you would be restricted in what you were able to produce; similarly, if your toolkit had a state-of-the-art machine that could screw, drill, glue and saw, but there were no instructions for its use, you may struggle to master or perfect its operation.

A toolkit in its simplest form is a container or vessel into which we place the tools. It is not the toolkit that is important but the tools we choose to place therein. The tools we place are influenced by the knowledge we have. Knowledge of a tool does not necessarily mean that one knows how to use it. Far too often, I have witnessed students loading their toolkits with resources they do not yet understand, which have limited applicability to the subject matter being studied, and contain information that requires a thorough understanding of core themes and concepts. It is difficult to build and then use an anatomical toolkit if one does not know what to place in it. I encourage my students to start with the basics, which include, a good resource text, ability to understand the content being studied, the innovation to develop learning aids such as cue cards, the inspiration to plan and locate areas for learning, and time and effort to invest in the learning. If you have a hammer, use it on a nail, not on a leaking pipe.

What Happens When You Use the Wrong Tool?

Consider the situation where you are asked to change a flat tyre. The only tool you have is a hammer. You may be able to hammer the tyre but the consequence of this action would be further damage, incurring greater expense, and it could severely retard the function of the vehicle. In anatomy, a similar situation may arise in terms of learning and/or mastering different or new knowledge. Let us use the example of studying the skeleton. Typically students would expect assessment questions to reflect the content they have learned. In this instance, the tool may be understanding the organisation and function of the skeleton, which could be used to address questions that ask for organisational components. If the assessment question asked for an understanding of how muscles produce movement, simply knowing or using knowledge of the skeleton would be insufficient to fully address the question. This could negatively impact the learner's ability to provide a meaningful and appropriate answer to the question. The result may be a low grade. In this example, the student may wish to focus on what he or she knows, and attempt to write around the subject matter without addressing the question or problem being asked. With the changing of the tyre, if we simply hammer the tyre, we may not recognise the need to identify or locate different and additional tools to complete the task. It is this recognition and application of different or new thinking that enables students to transform their learning and application of subject matter. It further provides a monitor for what is known and what needs to be known or learned.

The principle behind the toolkit is to promote and engage the learner with new and different ways of managing and conceptualising the study of the human body. It is not only what we put into our toolkits that determines our understanding, but what we take out to fine tune and develop further understanding. Through the metaphor of the toolkit we learn to know what we know and what we don't know. It is the recognition of what we know that drives forward the mechanism and motivation to seek out what we don't know.

It is essential for teachers to recognise the tools their students use, to promote opportunities to use these tools, to demonstrate different tools and to develop alternative uses for the tools. Once learners appreciate how to use their tools to construct and deconstruct knowledge, to build bridges between concepts and ideas, and to examine interrelationships, learning takes on a new and often heightened experience, transforming questioning into inquiry and laying the foundations for learning with a purpose.

To help shape the anatomical toolkit, the following questions may be useful:

- What anatomical toolkit do you have?
- What anatomical toolkit do you need?
- Which tools do you use?
- Which tools don't you use?
- What training or learning experience do you need to operate the tools more effectively?
- What is the value of having the basic tools and what tasks could you accomplish?
- What is the versatility of the tools and what are their different applications?
- Which tool do you use the most?
- Which tools do you use incorrectly?
- Which tools do you need to develop?

The tools we use or choose not to use may not be physical in nature. The anatomical toolkit is a collection of tools and skills that enable learning and creation of knowledge.

Figure 8.2 is a schematic illustration of the reshaping of the anatomical toolkit, filled with, and enhanced by, different skills. Think about these skills and whether they apply to your individual learning and/or teaching approaches. You may decide to rethink which tools and skills should be in the toolkit, but equally to develop those you have and use. The shape of the toolkit will change and continue to develop, provided you understand the principles behind the toolkit and the value that the tools and skills play in promoting and shaping learning and knowledge development and exchange. It is only by having a space to place our tools that our tools have a place within our learning space.

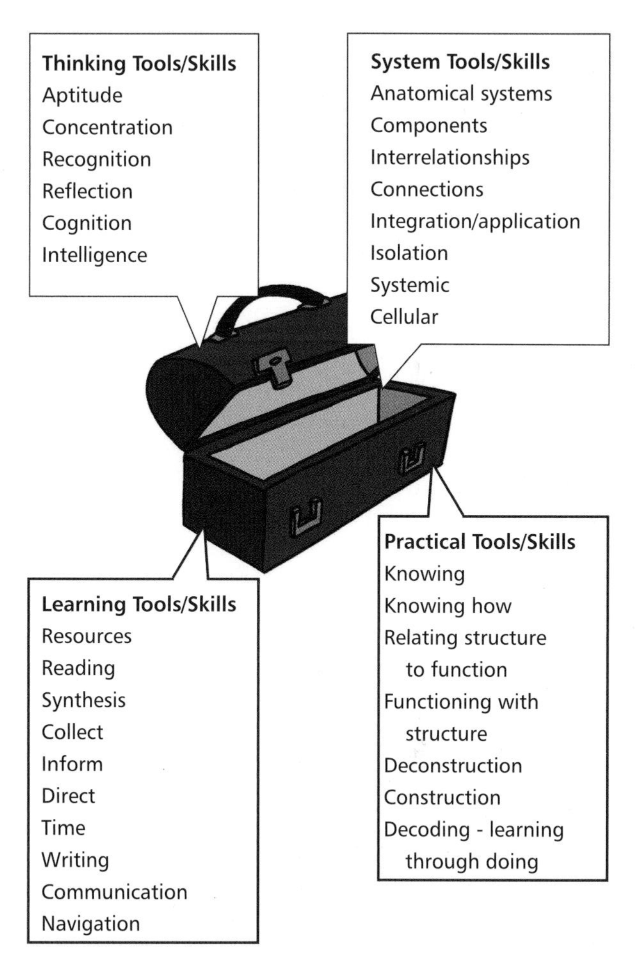

Figure 8.2: A schematic illustration of the reshaping of the anatomical toolkit, filled with, and enhanced by, different skills.

Towards Reflection – Learning to Own Anatomical and Physiological Knowledge

> *It is not sufficient simply to have an experience in order to learn. Without reflecting upon this experience it may quickly be forgotten, or its learning potential lost. It is from the feelings and thoughts emerging from this reflection that generalisations or concepts can be generated. And it is generalisations that allow new situations to be tackled effectively.*
>
> Gibbs (1988)

The study of anatomy and physiology, as reflected and depicted within the chapters of this book, provides a platform for considering and exploring the living body from its micro- to its macrostructures. It is through this on-going and sustained journey that we, a community of students, learners, teachers, educators and academics, begin to develop methodologies for learning and practice and learn to question what we believe to be true. How does this evolution occur? How do we travel from knowledge to knowing, and what signposts enable us to recognise what it is we truly know? This part of the chapter discusses reflective learning as a key component of knowing and provides guidance on how best to develop reflective practice for both the learning and the teaching of anatomical and physiological subject matter. The chapter concludes where the book begins, and explores the concept and creation of the anatomical toolkit, a workable space to store and build knowledge.

> *By three methods we may learn wisdom: First, by reflection, which is noblest; Second, by imitation, which is easiest; and third, by experience, which is the bitterest.*
>
> Confucius

> *Reflection is what allows us to learn from our experiences: it is an assessment of where we have been and where we want to go next.*
>
> Kenneth Wolf

Looking In to Find Out – Understanding Reflective Learning

What is reflection and how best do we engage students in the process of thinking about thinking and doing – think about it?

Many researchers have considered what reflection is or may become, and have tried to simplify a rather complex emotional and cognitive state. Moon (1999), for example, defined reflection as a set of abilities and skills, to indicate the taking of a stance and orientation to problem solving or state of mind. Boud et al. (1985) considered reflection to be an important human activity in which individuals recapture their experience, think about it, mull it over and evaluate it. They considered working with experience to be the primary objective for learning success. This understanding of reflection and reflective learning was further corroborated by Tate and Sills (2004), who made clear that experience and exploration of that experience provide a lens to view knowledge and knowing in different ways.

Reflective learning has two parallel processes; namely, informal and formal reflection.

Informal Reflection	Formal Reflection
Involves self-questioning	Draws upon research and evidence-based learning
Develops awareness of one's own assumptions	Provides guidance and frameworks for practice

What is evident is that reflection involves a cyclical process of questioning, re-questioning, and developing skills to appreciate and recognise the questions we ask, which may well be rhetoric at times. Having this brief understanding of what reflection may be or mean, how does it help in learning and developing anatomical and physiological knowledge – simply put, why is there a necessity to reflect?

Reflection helps learners to:

1. Understand what it is they already know – confirmation of prior *individual* learning.

2. Identify what it is they need to know in order to further develop subject matter (contextualising problems and solutions).

3. Make sense of new information and feedback in the context of their own experience, i.e. making their own experiences more meaningful (relational).

4. Guide and shape choices for further learning (providing developmental opportunities to enrich knowledge through application).

Conceptually, the four points above provide key markers and milestones within the learning cycle framework below.

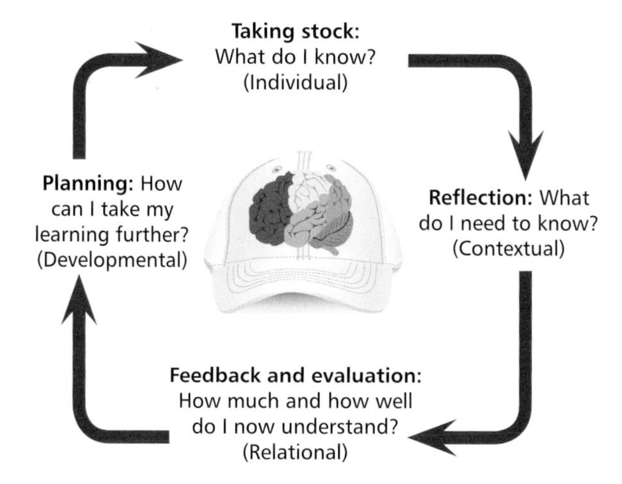

Figure 8.3: A learning cycle framework.

Dewey (1933), in his seminal work on thinking and reflection, developed a conceptual framework to assess what it is we need to do to develop reflective-learning skills. Dewey identified five stages that are fundamental for developing knowledge through knowing:

1. In the first stage, he outlined that we identify a problem or learning cluster that is perplexing and felt. This could be a bottleneck or threshold concept as discussed in Chapter 5 – Coming Unstuck.

2. The next stage involves refinement and observation of the identified problem to create a deeper understanding of the component parts.

3. In the third stage we begin to develop a hypothesis or understanding about the problem, its origins and possible solutions.

4. The fourth stage is where we subject the hypothesis to scrutiny and reasoning.

5. Finally in the fifth stage we test the hypothesis or understanding in practice.

This early conceptualisation of reflective learning in practice enabled scholars to consider the key components of reflective practice and further development frameworks for thinking *about* action (i.e. reflecting on what has happened) and thinking *in* action (i.e. while the action is taking place). Schon (1983) wrote about reflection in action as a process that concerns thinking about something whilst engaged in doing it; having a feeling about something and practising according to that feeling. Kolb (1984) identified critical learning dimensions. The first is the concrete experience – the event. The event acts as a trigger to initiate a learning cycle, followed by a process of reflective observation. During this process the individual learns to consider what has happened from a variety of perspectives, appreciating personal and group feelings and actions. The next dimension is abstract conceptualisation – the use of theory to repackage reflective-learning ideas. Finally, through active experimentation the student, armed with new or renewed understanding, attempts to revisit a learning situation differently.

Figure 8.4 depicts the overlap between Kolb's stages and considers the dynamics of student–teacher reflective cycles.

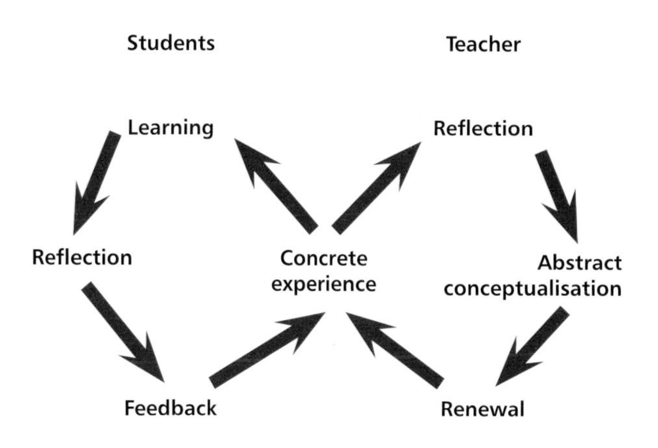

Figure 8.4: Integrating Kolb's stages of reflection.

In the study of anatomy and physiology, there are numerous opportunities to stop and think about what it is that we are experiencing whilst doing a specific task. Question sequencing is a useful learning aid to develop reflective practice learning in action.

It is important to note that whilst Chapter 5 – Coming Unstuck focused predominantly on difficult learning situations, the content within this chapter can be used to further emphasise and recognise what it is that needs to be learned to navigate difficult subject matter. This chapter should be read in conjunction with the one on difficult concepts.

Johns (2000) identified ten Cs of reflection, which provide an interesting scaffold for dealing with learning and knowing in different ways. His model, outlined in the table below, considers a continuum of personal action and development.

In developing a framework for reflective learning, theorists have tackled the questioning of action planning. In order to know, we need to know how to know by understanding the components and connections required for self-direction and knowledge mastery. Gibbs's reflective cycle (Figure 8.5) illustrates this process and uses a development framework to embed critical questioning into the learning dynamic.

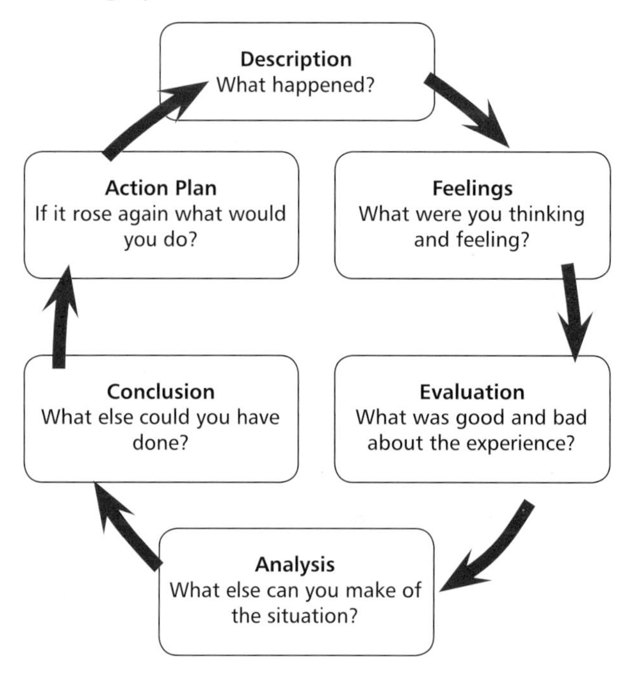

Figure 8.5: Gibbs's reflective cycle.

Commitment	Accept responsibility and be open to change and learning structures
Contradiction	Note and document tensions between actual and desired learning
Conflict	Harness energy to take action by focusing on learning and not necessarily avoidance of learning
Challenge	Confront actions, beliefs, barriers, attitudes and boundaries in a non-threatening way
Catharsis	Work through negative feelings
Creation	Move beyond old self to new and novel alternatives – learn to realise potential
Connection	Connect new insights into learning and teaching
Caring	Nurture desirable learning/teaching characteristics
Congruence	Reflection as a mirror for caring
Constructing	Build personal knowledge in practice and learning

Ten Cs of Reflection (Johns [2000]).

It is all very well having theories and frameworks for reflective learning, but in practice how does this work? Each individual will travel a different and subjective journey in realising the benefits of and barriers to reflection. What is the impact of reflective learning on knowing and the construction of a knowledge toolkit equipped with the tools necessary to deal with the challenges of learning that lie ahead?

Barriers and Benefits

Personal reflection incorporates the following:

- *Priorities* – What is important now compared to what becomes important in the future. Often students struggle with prioritising their learning and leave critical learning to the end. Learning to learn is about perfecting priorities and creating learning spaces to digest learning concepts.
- *Time management* – Finding time to learn is fundamental. One needs to balance quality and quantity of learning. It is important to find time but more important to use the time you find effectively.
- *Motivation* – The more motivated a learner is or becomes, the more inspiring the learning time will be. Intrinsic motivation enables students to develop internal strategies to inspire learning.
- *Direction* – Reflective learning must have direction and purpose and ask questions about the intent and outcome of the learning process.
- *Strengths/weaknesses* – Identify strengths and weaknesses in learning and reflection and use the strengths to manage the weaknesses.
- *Control* – Looking inwards and appreciating that students are in control of their learning mitigates against blaming external cues for failures or lack of learning.
- *Action plan* – Learners need an action plan to drive their learning goals. Planning, and planning shrewdly, creates order and an opportunity to consider learning and reflection in a synchronised way.

Practically, reflection involves a number of time-sensitive and emotionally influenced steps:

- Slowing down
- Pausing to examine, analyse, inquire and appreciate the complexities of life as well as the subject matter being studied
- Making the active choice to pause and examine
- The cognitive processes of analysis, synthesis and evaluation

Reflection has many different forms and functions – there are multiple ways to reflect, connect and refine.

Moon (2004) provides a useful platform to explore and consider the emotional pitfalls inherent in reflective learning by engaging the concept of emotional intelligence:

- *As part of the knowledge which is involved in reflection* – e.g. 'I know that I feel uneasy about this situation and therefore I will need to take it carefully and manage my feelings.'
- *As involved in the process of reflecting* – e.g. a feeling of unhappiness will tend to colour the manner which we reflect.
- *As an outcome of reflection, arising from the process of reflecting* – e.g. I reflect on how it was when I was on holiday in a wonderful place and, as a result, feel more positive and optimistic about things.
- *As a potential inhibitor or facilitator of reflection* – e.g. I am feeling as if everything is flowing along nicely, and this helps me to reflect on my behaviour around my difficulty in learning more effectively.
- *As a trigger for reflection*

- *As the subject matter of reflective learning, 'emotional insight'* – e.g. After an intimate talk with a friend, I somehow seem to see a difficult learning situation more clearly and can handle it better. I am not conscious of learning anything in particular.

Moon (2004, 2006) continues to explore how reflection influences learning and how learning becomes embedded in reflective practice. She considers how reflection might enhance learning through reflective teaching:

1. *Reflection is part of learning* – It seems that reflection is involved in meaningful learning where the learner is seeking to make sense of new material, linking it to what is already known and, if necessary, modifying prior knowledge and understandings to accommodate the new ideas.
2. *Reflection is associated with situations where there is no new material of learning* – This is where we make sense of ideas (knowledge and understandings) that we have already learnt.
3. *We learn from the representation of learning* – When we represent learning in writing (for example), in a sense it becomes new material of learning and we can reinforce the learning or check our understanding of it, using it as a feedback system.
4. *Reflection also facilitates learning by enhancing the conditions that seem to favour learning* – These include (all cited in Moon 1999):
    a. The provision of 'intellectual space' (Barnett, 1997) – Reflection slows the pace of learning.
    b. The development of a sense of ownership of learning – This has long been recognised as an important basis of learning (Rogers, 1969).
    c. The development and improvement of the process of learning to learn – Students who achieve well are more often students who are aware of their own learning processes – their weaknesses and strengths (Ertmer and Newby, 1996).
5. *Reflection engages the learner with the 'messiness' of learning as a process* – The learner learns to cope with ill-structured material of learning.

How Do We Use Reflective Learning to Develop Knowing?

This question is complex and involves a delicate understanding of various components of knowledge, knowing and knowledge application. It is difficult to conceptualise how readers of this chapter will use the content to refine and/or define their thinking in terms of anatomical learning. Below is an attempt to present ideas, provide strategies for thinking through questions, and position reflective learning within a paradigm of, and for, change.

For me personally, reflective learning is a state of mind, influenced by learning styles and teaching methodologies and pedagogies. It is important to recognise that not all students are comfortable with reflective learning and writing, as some may be action oriented. Learners may need guidance in learning how to reflect and understanding the impact of reflection on their learning. Moon (2004, 2006) provides some useful triggers to help teachers better prepare and support students with their reflective-learning development:

1. Consider why reflection is being used to facilitate this area of learning.
2. Consider how reflection differs from more familiar forms of learning.
3. Give examples of reflective learning/ writing – good and poor.
4. Generate discussion of students' conceptions of reflection.
5. Enable practice on reflective writing and provide opportunities for feedback.
6. Give a starting exercise that eliminates the blank page.
7. Have other tools available to help students to get started.
8. Expect to support some students more than others.

These triggers are bounded by the following:

- Establishing objectives
- Recognising current level of performance
- Planning how to meet targets
- Using effective time management
- Using feedback and support
- Monitoring and reviewing progress
- Critically reflecting on own learning.

By engaging the process and principles of reflective practice we learn to build bridges to reflection to further ensure and secure:

- *Non-judgemental support*, e.g. using mentors and teachers to help support student learning of anatomy and physiology.
- *Feeling 'safe' enough* – to do more than simply passing an assessment. This is about building confidence in your ability to enjoy the learning of anatomy and physiology.
- *A role model*, e.g. a mentor who reflects on their own learning and is able to disseminate good learning practices such as making notes and annotations, using a range of anatomical resources, visiting museums and dissection halls, or watching educational videos.
- *Knowledge of as many methods as possible*. This means we need to consider as many different ways of learning or seeing anatomy and physiology as possible. By exploring multiple methods we may be able to make better sense of how structures relate.
- *As many opportunities as possible for engaging in reflection*, e.g. pairs, groups.
- *Time and energy* – setting aside time and energy to study.

The greatest bridge to reflective learning is being able to communicate the storyline within the reflective-learning process. Knowing involves a personal interpretation of learning content explored through narratives. Narratives invite the student to develop mental rehearsal scripts and practise the dialogue and messages within the reflective-learning process. This enables connections between personal and professional development and sharing of knowledge to promote deeper reflection and comparisons.

Strategies for Promoting Reflective Learning

1. Reflective journals
2. Sequential questioning (using the learning of the sliding filament theory as an example):
    a. Reactions to a learning experience. How did learning the sliding filament theory make you feel?
    b. Different ways to see a problem. Is what you learned about the sliding filament theory true for all processes within the body?
    c. How this particular learning experience compares with other experiences. How did learning the sliding filament theory compare with circulation of blood through the heart?
    d. How has your experienced been enhanced or inhibited by prior knowledge?
    e. What have you learned about your learning and understanding of the sliding filament theory?
    f. What do you still need to learn and why?
    g. What has prevented your learning of the sliding filament theory (lack of knowledge, difficult conceptualisation, lack of meaning, lack of knowing how the process becomes useful)?

h. How might you address or achieve your identified learning goals through your action plan?

i. How will you use your new learning to change your learning outcome and direction? By knowing the components of the sliding filament theory, how has that influenced your understanding of anatomical processes?

Through these questions, one may become more conscious of one's potential for bias and discrimination, making the best use of available knowledge, avoiding past mistakes, and maximising opportunities for learning.

Reflective learning is an intricate process filled with barriers and benefits. Barriers to reflective learning may stem from fear of judgement or criticism, being closed to feedback and feedforward, defensiveness, and arrogance. Reflective learning is an active process, one which requires and demands time and effort. To effectively reflect and learn from and through reflection, one should actively set aside time for this as part of the working day.

The responsibility rests with the learner and teacher to recognise the importance and purpose of asking questions and generating insight into personal and professional development. Unless we make conscious and systematic efforts to critique our own learning and learning practices:

- We will be unaware of how and when we are being discriminatory.
- We will not make use of the knowledge base developed in the subject matter and influenced by the profession.
- We will continue to repeat the same mistakes.
- Our skills will stagnate rather than develop.

Reflective learning enables students to set the standards for their own learning whilst working towards learning attainment. This provides a progressive marker for personal success and/or failure. Having endured the process of writing, in part, the content for this book, I fully appreciate the rigorous and often frustrating process of reflection. I realise the hours spent reading, rereading, writing, revising, refining, deleting and rethinking the chapters, the content and the flow of the text. I believe that reflection is a lived experience, as it enables one to visualise a process by developing goals and milestones to celebrate achievement and calibrate progress. By learning to reflect and distancing myself from disturbance, I am better able to explore my experiences more fully, draw upon influential occurrences, and learn to capture in words what I experienced in my own space and time. For me, writing this book is the start of the reflective-learning process, not the end; it is the enabler for concentrating on what it is I wish to achieve and then carefully asking questions in relation to my set objectives. Reflection is learning about how one copes with being oneself.

Harry stared at the stone basin. The contents had returned to their original silvery white state, swirling and rippling beneath his gaze.

'What is it?' Harry asked shakily.

'This? It is called a Pensieve,' said Dumbledore. 'I sometimes find, and I am sure that you know the feeling, that I simply have too many thoughts and memories crammed into my mind.'

'Er,' said Harry, who couldn't truthfully say that he had ever felt anything of the sort.

'At these times,' said Dumbledore, indicating the stone basin, 'I use the Pensieve. One simply siphons the excess thoughts from one's mind, pours them into the basin, and examines them at one's leisure. It becomes easier to spot patterns and links, you understand, when they are in this form.'

J.K. Rowling, 2000, *Harry Potter and the Goblet of Fire*

In the Beginning ...

'In the Beginning', the opening words to the most famous and widely read book in the world, the Bible, we learn about the creation of life. We learn about the process of life and the formation of living

things. Over the years, science has revolutionised our understanding of how things came to be and what *being* means. Through science we have learned to dissect the human body into its basic units of life and then produce new life outside of the body. Science has encouraged transformation of thinking, development of ideas, reconfiguration of all things living, redevelopment of body systems, and repair of dysfunctional and diseased body systems. Over time we have learned that the only known constant is change itself, and through evolution we learn to question the basis for life and the development and shape of future lives. From birth to death, the human body epitomises growth and development which no machine can truly replicate. The formula for intellectual life is filed within the cortex of the brain and kept as man's best secret. Over time we learn to recognise the pieces of this life puzzle we call the human body.

The study of anatomy and physiology enables us to study each component and then attempt to piece them together into a functional three-dimensional body that is capable of thinking, moving, growing, developing, living and, finally, dying. The study of this body has become the fascination of many. Just when we think we know, we learn we don't really understand. This paradox of learning and unlearning punctuates the study and complexity of the body. Just as the body has multiple interconnected layers, so too does the study of anatomy and physiology. The Bible relays stories of creation; so too do we create stories about the events and processes unique to the body.

In the Beginning we considered what it is like to speak anatomy; to develop useful study strategies and techniques; to devise relevant assessment tools and tasks; to navigate conceptually difficult subject matter; to think about seeing, feeling and doing anatomy as a lived experience; and finally to reflect upon these experiences by building anatomical toolkits and exploring resources. The anatomy puzzle is not simply about solution, it is a canvas to explore the shape and form of the individual pieces and examine how best they fit and work together. It is about how not fitting together opens new lines of inquiry and knowledge growth. It is about pushing boundaries and analysing the boundaries we push. Science has placed the body on film, on show, on slides and on stages to peel back the seen and explore the hidden. When we become aware of the unseen, the hidden, the unknown, shifts in our thinking drive forward different ways of studying and categorising body segments and systems. Through function we recognise dysfunction and engage with processes to discover answers to defects. Through internal invasion, we build defence mechanisms and immunity. The human body has taught us that nothing is impossible and everything is possible, illustrated through human endurance, endeavour, engagement and exploration of functional elements in structural matrices. The body has been used as a design template for inanimate objects such as bridges and arches. We are now beginning to appreciate the roles of and interconnections between muscle and fascia, and realise that body strength is a composite of multiple design elements.

Despite numerous attempts by designers, architects and artists to recreate the dynamics of the human body, few have succeeded in truly appreciating the intricacies of the body. I remember one of my professors explaining how a metal worker wanted to construct the muscular system using metal offcuts and filings. He spent three years researching and examining the muscular system, calculating weight and dimensions of each muscle. He painstakingly hand-cut each muscle and carefully positioned the muscles on to a metal skeleton. Once he had completed the assembly of his metallic muscular system, he stood back and was amazed at how this metallic body balanced in perfect harmony. We take for granted the dynamics of the body, the balance and association of structures, and often overlook the alignment of structures. Perhaps the most important skill we need is our ability to observe what it is we have.

In the Beginning we observed, we asked, we questioned, we learned to study – now we know that we do not know, and have revised the way we study so that questioning leads to inquiry, leads to scholarship, leads to dissemination, leads to explanation, leads to further study. *In the Beginning* is simply a point in time which may not have been the beginning, but what we have come to believe in as the origin of life.

Figure 8.6: The evolution of the modern revolution of knowledge.

Final Thoughts

As we close this chapter, we look back to look forward. We realise what it is we need to do, to learn, to develop, to question in order to build a deeper appreciation for what it is we know and have yet to learn. The next time you study or teach anatomy and physiology, ask yourself what it is that you believe is true about the content you uncover. Look for the evidence to support what you are studying and teaching. Consider alternative perspectives and challenge yourself to think beyond the content of the chapter, module or resource being used. Think about what it is you truly need to think about, so that the learning becomes the learned and the teaching the taught.

Anatomy and physiology are the vehicle for helping us uncover what lies beneath and exposing what we may choose to hide. By knowing this, we learn to know what knowledge building can become. We thus learn to make sense of the subject matter through connections and reflections.

References

Barnett, C. (1997). 'Sing along with the common people': politics, postcolonialism, and other figures. *Environment and Planning D: Society and Space* **15(2)**: 137–154.

Boud, D., Keogh, R. and Walker, D. (1985). *Reflection: Turning Experience into Learning*, Kogan Page.

Dewey, J. (1933). *How We Think*, DC Heath.

Ertmer, P.A. and Newby, T. (1996). The expert learner: strategic, self regulated, and reflective. *Instructional Science* **24**: 1–24.

Johns, C. (2000). *Becoming a Reflective Practitioner*, Blackwell.

Gibbs, G. (1988). *Learning by Doing: A Guide to Teaching and Learning Methods*, Further Educational Unit, Oxford Polytechnic.

Kolb, D. (1984). *Experiential Learning as the Science of Learning and Development*, Prentice Hall.

Moon, J. (1999). *Reflection in Learning and Professional Development*, Routledge.

Moon, J. (2004). *A Handbook of Reflective and Experiential Learning*, Routledge.

Moon, J. (2006). *Learning Journals: A Handbook for Reflective Practice and Professional Development*, Routledge.

Rogers, C.R. (1969) *Freedom to Learn*. Cited in Panarchy (2000–2016). *Carl R. Rogers, Freedom to Learn (1969)*. Available online at http://panarchy.org/rogers/learning.html. (Accessed 15 August 2015.)

Schon, D. (1983). *The Reflective Practitioner*, Temple Smith.

Tate, S. and Sills, M. (eds) (2004). *The Development of Critical Reflection in the Health Professions*, Higher Education Academy.

9 Resource Anatomy

Throughout this book we have mentioned teaching and learning techniques, games, props and websites that can be used to demonstrate key anatomical points that will help the student to understand the topic and therefore have a better chance of retaining the information. This chapter gives more details of these techniques and will allow you to form a 'toolkit' to help you build your knowledge. Students can utilise these tools to create memorable learning strategies and to affirm a deep understanding in anatomy and physiology. Teachers can use these tools to vary their teaching styles to match the students' learning needs, and to lighten the atmosphere within the classroom.

As with any toolkit, how you use the tools will depend on your needs. Explore how these ideas can be changed, amended and improved to suit your own needs and the subject being learned, and can be adapted for different topics. Be creative! And do let us know what you have invented.

Each of the senses has been examined in previous chapters, giving practical suggestions for different learning and teaching strategies to suit all learners' needs. This chapter offers practical ideas for learning, both for the student and for the teacher, to promote learning.

Pen and Paper Required

Diagrams with Text Boxes and Arrows

Some anatomy students are gifted artists, and are able to draw amazing pictures of bones, muscles, the digestive system, and so on. However, most anatomy and physiology students are concerned that their artistic flair is not good enough to draw their own diagrams. Anatomy teachers are completely aware of this, and will expect rather stylised diagrams rather than works of art. Taking time and effort in drawing your own diagrams will reap rewards as it will help visual learners to picture the details, and will help kinaesthetic learners to feel the three dimensionality of the structure.

The student will need to be able to identify the key structures in the diagram, and be able to label or annotate them. Consider printing and laminating the diagram with large empty text boxes attached to arrows indicating the structures that are to be identified. This allows white-board markers to be used to practise labelling the structures, which can then be wiped clean, enabling retesting another day. Also consider drawing diagrams on flash cards, with the details on the reverse side.

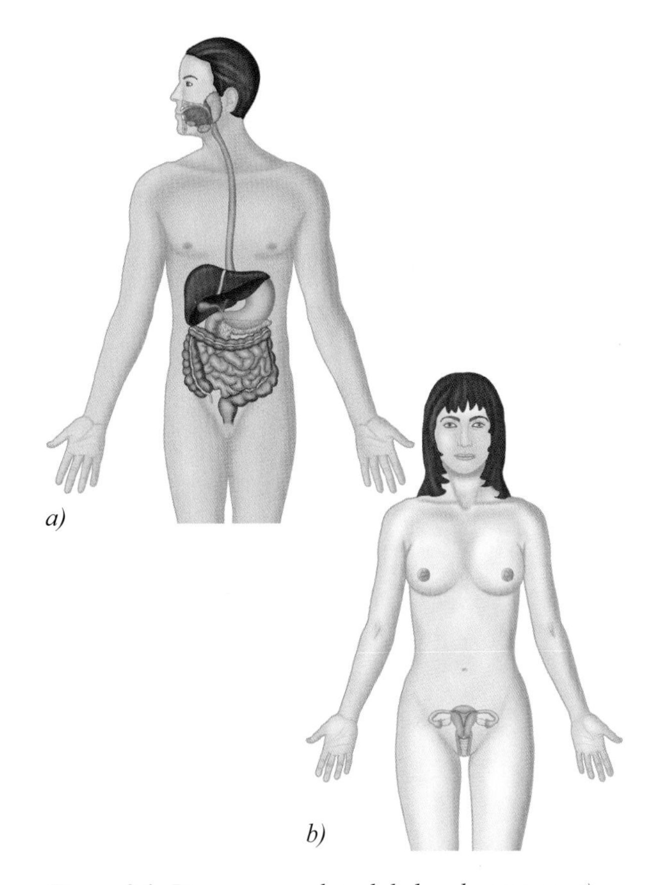

Figure 9.1: Diagrams ready to label and annotate, a) the digestive system, and, b) the female reproductive system.

Making Pathways

Many physiological processes follow defined pathways which need to be committed to memory. Write the names of pathways of physiological processes on cards and ask students to rearrange them into the correct order. This works well for the following systems:

Digestive System

Lips	Mouth
Pharynx	Epiglottis
Oesophagus	Cardiac sphincter
Stomach	Pyloric sphincter
Duodenum	Common bile duct
Pancreatic duct	Jejunum
Ileum	Ileocaecal valve
Caecum	Appendix
Ascending colon	Transverse colon
Descending colon	Sigmoid colon
Rectum	Anus

Renal System

These can be rearranged depending on which item is passing through the kidney, whether it be water, glucose, urea, protein or blood.

Nephron	Glomerulus
Bowman's capsule	Proximal convoluted tubule
Loop of Henle	Distal convoluted tubule
Collecting duct	Minor calyx
Major calyx	Ureter
Bladder	Urethra
Urethral orifice	Renal artery
Segmental arteries	Interlobar arteries
Arcuate arteries	Afferent arterioles
Interlobular arteries	Efferent arterioles
Glomerular capillaries	Vasa recta
Peritubular capillaries	Arcuate veins
Interlobular veins	Interlobar veins
Segmental veins	Renal veins

Cardiac Cycle

Inferior vena cava	Right atrium
Right atrioventricular (tricuspid) valve	Right ventricle
Pulmonary (semilunar) valve	Pulmonary artery
Lung	Pulmonary vein
Left atrium	Left atrioventricular (mitral) valve
Left ventricle	Aortic (semilunar) valve
Aorta	Systemic circulation

Building Models

Kinaesthetic learners will enjoy building models from household items to form visual analogies of body systems and structures.

Spinal Models

Although spines are held in shape by ligaments, a model of a spine can be made by threading cardboard tubes or cotton reels to model the flexibility of the segmented shape.

You will need:

- *Vertebrae* – 33 cotton reels, or small sections of cardboard or plastic tubes (from the inside of kitchen rolls or toilet rolls, or cut up drinking straws). The size of the sections of cardboard tube can be cut to represent 7 cervical, 12 thoracic, 5 large lumbar, 5 small sacral and 4 tiny coccygeal vertebrae.
- *Intervertebral discs* – 32 marshmallows or similar squashy sweets.
- *Spinal cord* – Embroidery thread.

Instructions:

Cut the sections of tube or straw into the correct vertebral body shape, using a diagram to give the correct representations. Using a large needle, thread the embroidery thread through each cardboard tube or cotton reel, interspersing with marshmallows, thereby representing vertebrae and intervertebral discs. Pull the thread tightly, holding the structure together firmly.

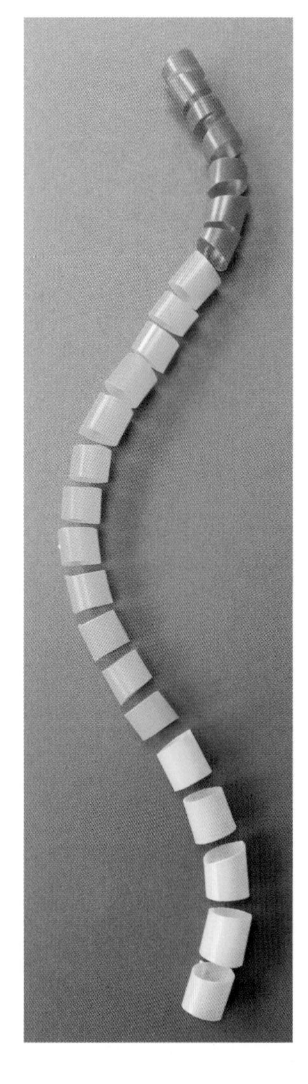
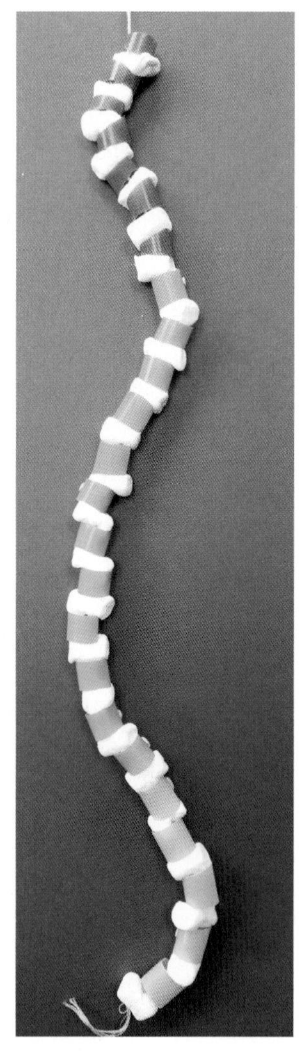

Figure 9.2: A straw forming the correct vertebral body shape. Note the flexibility of this structure, and how when the 'spine' is bent, the marshmallow discs bulge slightly.

Plumbing Pipes for Digestive System

A trip to the plumbing merchant's is a must to buy sufficient piping to form the tubes and flexures of the digestive system and is particularly useful to demonstrate the three dimensionality of the duodenum or the colon.

More elaborate models of the digestive system can be made from a variety of tubes, hosepipes and plastic bags.

Elastic Bands for Muscles

A simple working model of the elbow can be made from household items, demonstrating the action of muscles. Adding an additional muscle on the opposite side of the tube can represent antagonistic muscles.

You will need:

- *Muscles* – a large rubber band
- *Bones* – two paper towel tubes
- Thumb tack or drawing pin
- *Elbow* – tennis ball or similar ball, or a hinged section of an egg box.

Instructions:

1. Poke a hole into a paper towel tube about 5 cm from one end.
2. Cut the rubber band to make a long rubber band string.
3. Insert one end of the rubber band into the hole, and knot it once it's through in order to secure it into place.
4. Poke a hole in the second paper towel tube about 5 cm from one end.
5. Insert the other end of the rubber band into the hole, and knot it once it's through in order to secure it into place. This will act as the arm muscle.
6. Arrange the paper towel tubes end to end so that each hole is facing to the left. These will be the bones of the arm.
7. Insert a tennis ball or hinged section of an egg box in between the tubes. This will be the elbow.
8. Keeping the tubes on the tennis ball, move the tubes outward and inward to stretch out the rubber band muscle. This represents the movement of a muscle in a human arm. Note how when the elastic band is allowed to contract, the 'elbow' flexes.

Sliding Filament Theory

(From www.nuffieldfoundation.org/practical-biology/modelling-sliding-filament-hypothesis.)

In principle you could provide any materials for making models of muscle contraction. In practice, some materials are particularly useful for modelling changes in the position of the myosin head, and for modelling binding and bending. These include plasticine or Blu-Tack, pipe cleaners and bendy straws. To distinguish one kind of protein chain from another it is useful to have materials in a range of colours. Cardboard, sticky tape and marker pens are also useful.

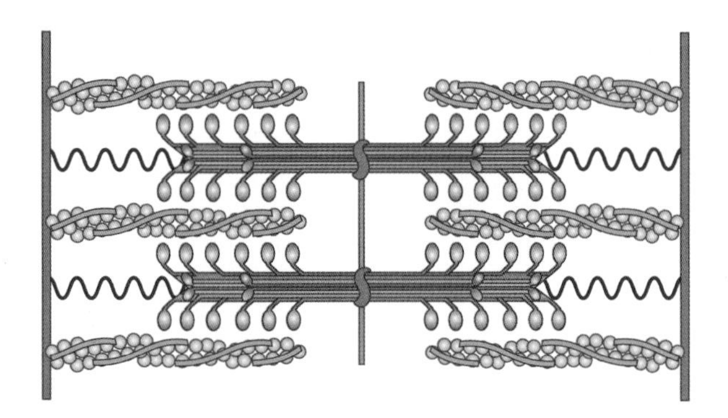

Figure 9.3: Sliding filament theory.

Some key things to look for in any animation or model of the sliding filament theory are:

- Explanation of the effect of calcium on the troponin/tropomyosin components – their change in shape exposes the myosin binding sites on the actin.
- Clarity of the sequence involving binding of ATP, conformational change in the myosin head, hydrolysis of ATP, and release of ADP and inorganic phosphate (P_i).
- As an ATP molecule binds to the myosin head, the head detaches from the actin. The ATPase on the myosin head hydrolyses the ATP, forming ADP and P_i (still bound to the myosin). This causes the change to the upright position of the myosin head. Binding and hydrolysing ATP puts energy into the system, and the head of the myosin molecule moves into the 'cocked'/right-angled position. When the myosin binds to the actin, ADP and P_i are released, and the myosin head returns to its relaxed position (or nods forward), causing relative movement of the filaments. Energy is transferred as work done by the muscles in contraction.
- Indication of several myosin heads on a single fibre acting one after another, rather than all myosin heads acting in concert.
- A link from the detailed molecular events to the overall changes in the sarcomere, and relative movements (and widths) of the H-band, Z-band and I-band.

Muscle/Fascia Microanatomy using Straws and Cling Film

Building a muscle from straws and cling film is an excellent way to give students an impression of the microanatomy of a muscle fascicle.

You will need:

- *Perimysium* – cling film
- *Epimysium* – cling film
- *Muscle fibres* – drinking straws.

Instructions:

1. Cut a piece of cling film 20 cm longer than the drinking straw, and 10 cm wide.
2. Place the drinking straw into the centre of the cling film and wrap it so that there is an overhang of cling film at either end.
3. Lightly twist the ends of the cling film to secure.
4. Repeat for 12 drinking straws.
5. Bundle these together to form a fascicle and wrap the whole fascicle in cling film.
6. Repeat until several fascicles have been constructed, then wrap them all together with cling film to form a muscle with tendons made from the combined cling film ends.

Figure 9.4: (a) Muscle/fascia microanatomy using straws and cling film; b) muscle fibres.

Peeling an Orange and its Segments – Fascia and Muscle Fibres

To create a visual analogy of muscle fascicles, an orange can be dissected: likening the orange peel to fascia, the pith to epimysium, and the segments to the muscle fascicles and their perimysium, each consisting of small segments of orange which can be likened to the individual muscle fibres.

Figure 9.5: Peeling an orange and its segments.

Jelly Moulds for Body Parts

Jelly moulds of various body parts can be purchased online. They are often to be found in Halloween or horror shops for obvious reasons. Moulds of hearts and brains are usually the easiest to source (see Amazon.co.uk). Red-coloured jelly will make the heart or brain more realistic, and adding some evaporated milk to it will make the pink colour more lifelike.

Jelly Moulds of Cells

You will need:

- Cell membrane – vessel shaped like the cell being modelled: round mixing bowl, lunch box, plastic bag
- Cytoplasm – gelatine mixture or a light-coloured jelly (orange or lemon is ideal)
- Nucleus with nucleolus – peach half or plum half with stone removed
- Mitochondria – raisins, liquorice comfits or mandarin orange slices
- Endoplasmic reticulum – jelly worms
- Rough endoplasmic reticulum – jelly worms dipped in sprinkles and dried before adding to jelly
- Golgi body – fruit ribbon, folded up
- Ribosomes – peppercorns
- Lysosomes – Smarties or M&Ms
- Vacuoles – gobstoppers or jelly sweets
- Microtubules – dried spaghetti or liquorice bootlaces.

Instructions:

Pour the cooling jelly mixture into the vessel, and position the organelles as indicated in the textbooks.

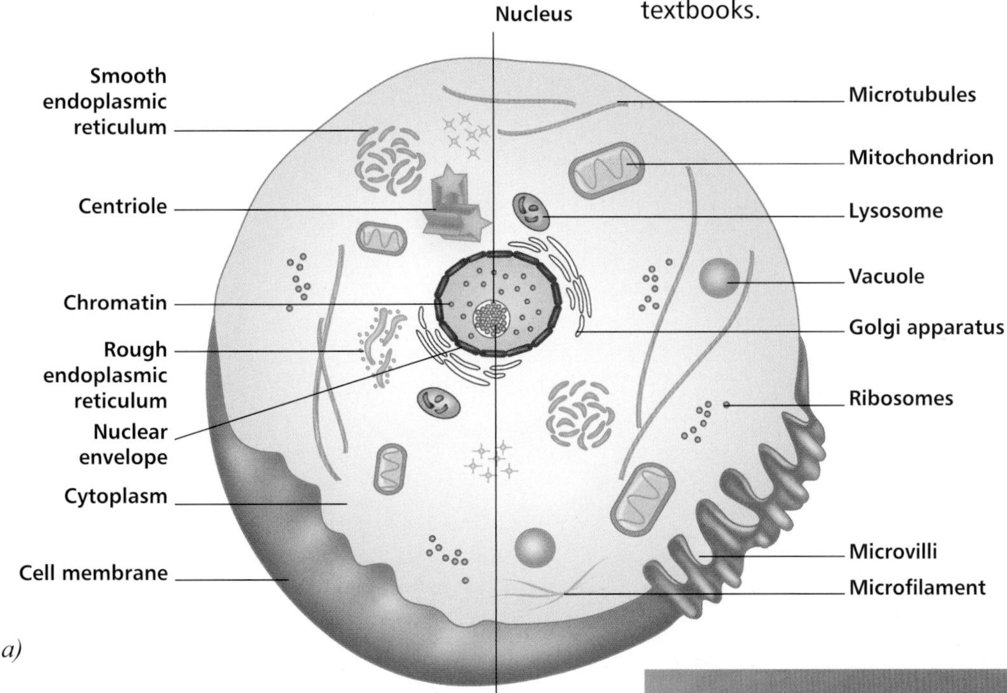

b)

Figure 9.6: a) Diagram of a generalised cell, b) jelly mixture with organelles arranged.

Masking Tape or Electrical Tape for Muscles on a Full-Sized Skeleton

Full-sized skeletons are a fabulous resource to help memorise the locations of muscles and ligaments. Not only can the bones be labelled with little stickers, but the muscles can be replicated with craft materials. Use coloured masking tape or electrical tape to demonstrate the origins and insertions, bellies, and tendons of muscles. For larger muscles, additional paper or card can be used. Smaller pieces of tape, or even Blu-Tack can be used to show the location of ligaments.

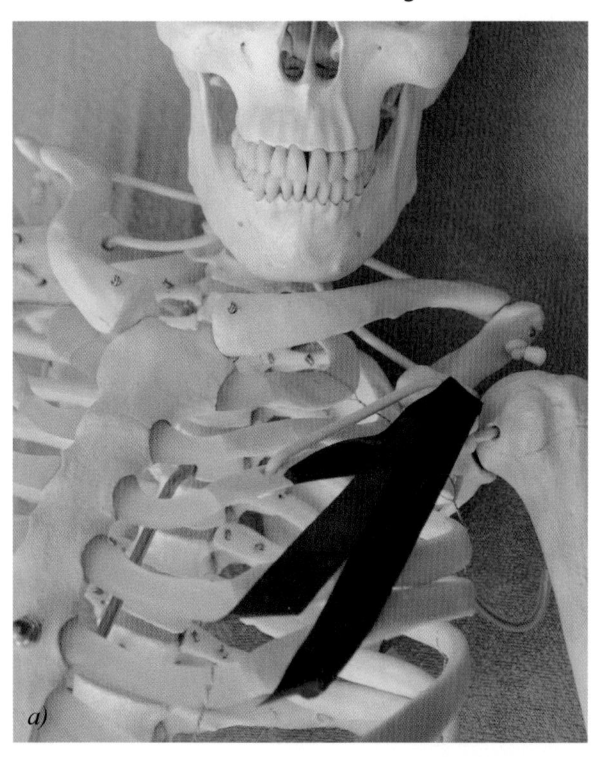

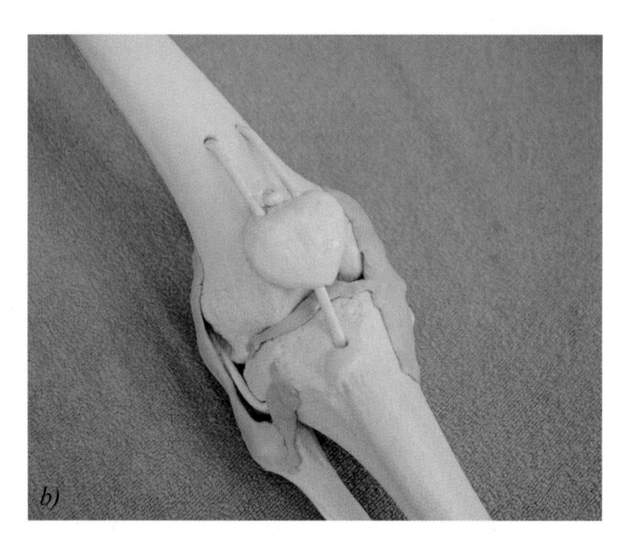

Figure 9.7: a) Masking-tape muscles; b) ligaments made of Blu-Tack.

Edible Cells Made in Fondant Icing

The creative chefs in the class will enjoy fashioning structures such as individual cells or even whole organs in coloured fondant icing. Small versions can be used to place on top of cupcakes and can be given as prizes in anatomy quizzes, whereas others can be freestanding sculptures.

Modelling Clay Organs

Another sculpting opportunity, but this one is less likely to induce diabetes! Coloured modelling clay or homemade salt dough can be used to fashion organs and structures. Encouraging discussion about the location, structure, function and detail whilst sculpting means that the students think carefully about the item being modelled.

Sculptures of Skin

The many layers of the epidermis and its underlying dermis and subdermis, filled with structures such as hair, nerves and pores and deep fatty layers, lend themselves to modelling using a variety of materials such as modelling clay, rice, pasta and pipe-cleaners, and even vegetables. Build the layers of the epidermis from ribbons or cling film, and use bubble wrap as the subdermal fat layers. Embed pieces of pasta to show structures such as hairs, pores or sebaceous glands. Use pipe cleaners to show blood vessels.

Collages of Organs and Cells

Use a variety of craft materials, card, paper or fabric to create detailed collages of organs, structures or cells. Always include a key so it is clear what each item represents. Encourage discussion about the shape and location of the structure in question, so that the students learn from each other, as well as from the information in the textbook.

Figure 9.8: Collage of a cross section of the skin.

Wool to Show Nerve Pathways or Circulatory System

For any anatomical item which shows a pathway, circuitry or vessels, use wool or string stuck onto an outline of the body or limb. Different-coloured thread can denote different vessels or paths. Annotate with details written on cards, then photograph the finished design.

Knitted and Stitched Structures

Not for the faint-hearted, but wonderful for the very talented knitter or seamstress. There are knitting patterns for some anatomical structures available on the internet. The time that goes into knitting a digestive system is immense, which means that the amount of time taken to build these probably outweighs the learning potential gained. However, if a keen knitter offers to construct such a masterpiece, accept willingly, as the knitted digestive system is such a talking point, and can be 'life sized', which means that the students will be able to consider the relative sizes and locations of its components, as well as being able to pick up the items and hold them without feeling squeamish.

Digestive Enzymes – Lock and Key

The specificity of an enzyme to its substrate molecule is likened to a lock and a key – only a key with the correct shape can fit into the lock; only one type of enzyme can speed up a specific reaction because of its shape fitting the substrate.

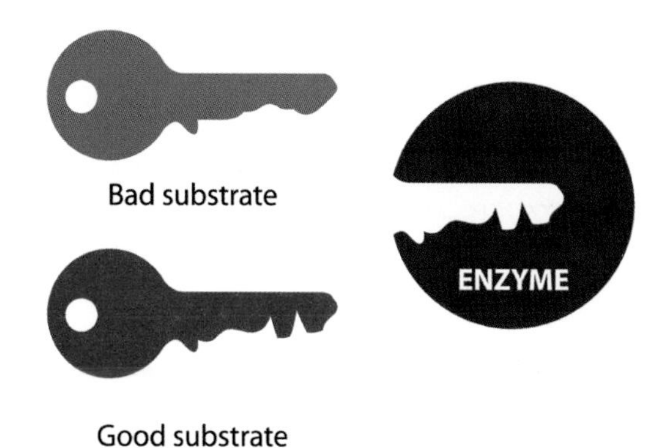

Figure 9.9: The specificity of an enzyme to its substrate molecule is likened to a lock and key.

Electrical Circuits for Nervous System

Students often misunderstand how the nervous system is constructed, and the way in which the autonomic nervous system is balanced by the actions of the sympathetic and parasympathetic systems throughout the body. By constructing a circuit from bulbs, cables, switches and a battery, a system can be constructed showing that the action taken in the body (i.e. the bulb switching on) occurs because of the balanced effect of the parasympathetic system (dimming the bulb) and the sympathetic system (brightening the bulb).

Structure of the Small Intestines Using a Microfibre Mop

The internal surface of the small intestines can be replicated visually and kinaesthetically by looking at the texture of a microfibre cleaning mop. The 'villi' are obvious as represented by the tufts, yet when you further examine the surface of the 'villi' you can see that there are 'microvilli' present in the form of the small microfibres on the tufts.

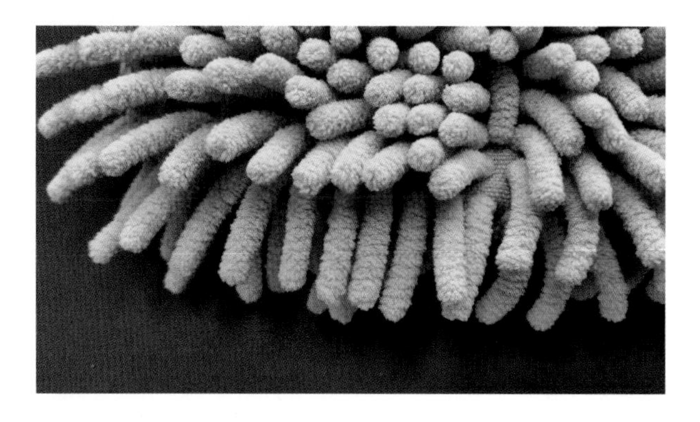

Figure 9.10: The tufts of a microfibre cleaning mop represent the villi, with the microfibre on the tufts representing the microvilli.

Brain Hats – Drawing on Peaked Caps with Brain Landmarks

External brain markings and key structures can be turned into baseball caps or swimming caps.

You will need:

- White swimming hat or plain white peaked hat
- Coloured felt-tipped pens.

Instructions:

The baseball hat can be drawn on so that the sections of the brain are clearly defined and labelled.

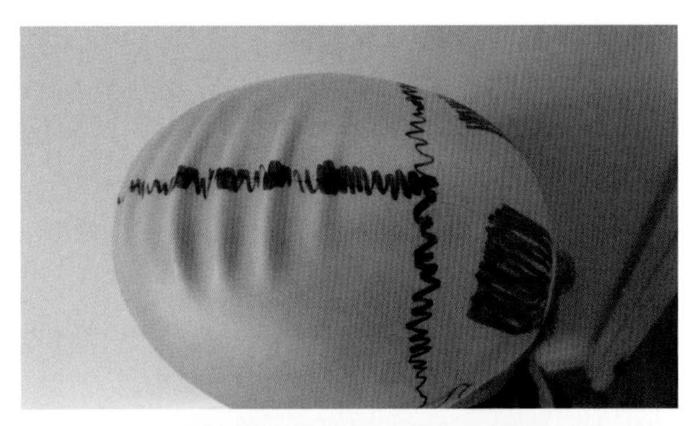

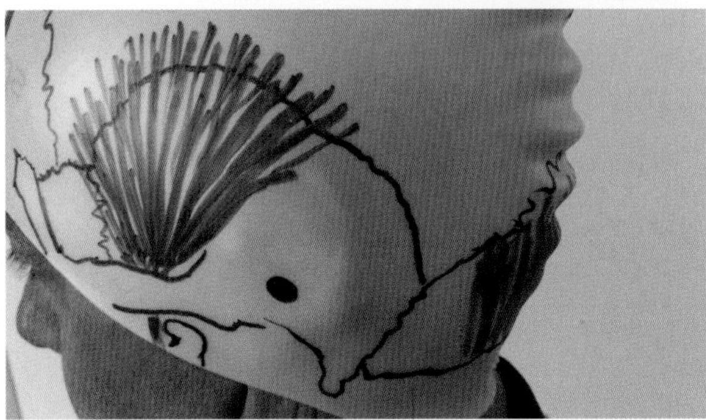

Figure 9.11: Swimming caps are best drawn upon while being worn, as they stretch; so find a willing volunteer to wear the swimming cap while you sketch on the structures.

Clothing Preprinted with Anatomical Diagrams

Fashion is a transient thing, with clothing following trends and designs, and in recent years there have been anatomical pictures printed onto T-shirts, gloves, socks, and even on tights. These can be utilised in the classroom and in study sessions with friends to recall and embed the anatomical positions of certain organs, albeit in a simplistic way.

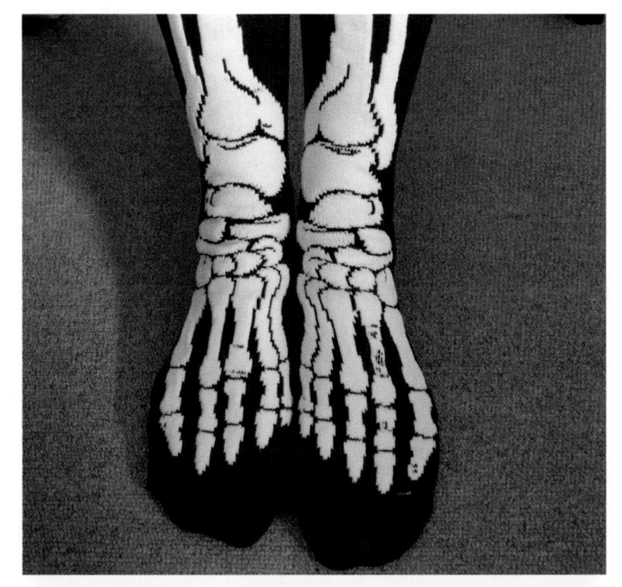

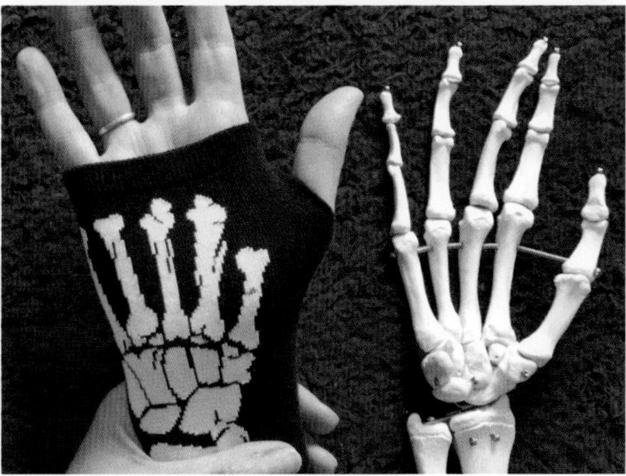

Figure 9.12: Anatomical clothing.

Hands-On Learning

Cornflour for Fascia and Palpation

When cornflour is mixed with water, the large cornflour particles remain 'suspended' (float around) in the liquid. Cornflour slime is thick because the particles are packed very close together, yet they are still able to slip past each other. When the mixture is slowly moved, it acts like a liquid because the suspended particles have time to move past each other. Yet, when sudden stress is exerted on the mixture, by rolling it or hitting it for example, the water quickly flows out of the area but the particles do not have enough time to move out of the way. The cornflour particles temporarily stay packed up where they are, which makes the slime act like a solid.

Although this non-Newtonian mixture is great fun to play with, and is often used as sensory play in schools and nurseries, it can be used to promote a discussion on force and reaction to force in the human body. Under pressure, our bodies will spread the load throughout the body, causing muscles to tighten and firm up. This can be palpated as tight tissue or muscle knots. Try hitting the cornflour mixture and note how it immediately stiffens. Stroke your hand through it slowly, and note how it turns to liquid, yet if you try to pull your hand from it too quickly, it grabs you!

Immaculate Dissection Courses

Attending courses that allow the student to see the body parts painted onto the body are invaluable for learning the exact locations of organs, muscles and landmarks. It gives a visual and kinaesthetic sense of location, and demonstrates the muscles during movement, so actions of the muscles can be visualised and learnt. See immaculatedissection.com.

Figure 9.13: Immaculate Dissection team at work. (Kathy Dooley, Anna Folckomer and artist Danny Quirk).

Analogy and Role Play

Acting out a physiological process or pretending to be a body part may seem to oversimplify anatomy and physiology; however, many students will learn huge amounts by taking part in such activities. The student will have to read up or learn the actions, locations and structure of the body part, and to work in a team to form a body system. Often a student will remember for the rest of their lives the names of their fellow students who were a sphenoid bone or an occiput.

Reverse-Engineering Discussion

One of the best ways of remembering how an organ or structure functions is to consider: if that body part were missing, what would happen to the body? It is best to make it as graphic as possible so that it can be visually recalled easily.

For instance, in the case of skin functions:

- Our bodies would be a sticky mess without an external covering.
- We would be open to infections with all the dirt sticking to our goo.
- With no melanin in our outer covering, we would be prone to shrivelling in the sun too – not a pretty sight.
- Just looking at our hairless gooey head will be a memorable image!
- And we would be shivering, as we would be unable to form goose-pimples to help us retain heat.
- Added to that glorious image, our knees will be bowed from rickets caused by the inability to synthesise vitamin D via our skin.

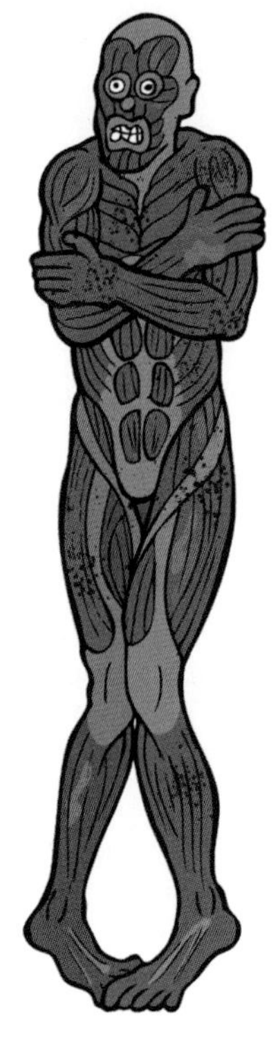

Figure 9.14: The body without the skin.

Think of the functions of bones in this way:

- Our bodies without bones would be a giant blob on the floor, like a massive amoeba.
- We would not be able to move, as our muscles would have nothing to pull on.
- We would need to deposit and store our calcium and magnesium and fat in other parts of our body – would we grow horns, or perhaps have some rocks attached to us as foundations?
- Where would blood be formed? A massive spleen and liver would be required for haematopoiesis, so our blobby bodies will have a giant bulge denoting the spleen and liver.

Figure 9.15: The body without bones.

Certainly a memorable image or two there!

Which Is the Most Important Body System?

A great discussion topic for a revision session is, 'Which is the most important body system?' or, for a given body system, 'Which is the most important structure in that body system?'

The majority of initial answers will probably focus on the heart and lungs as being the most important, as, after all, when they stop working we die. However, after discussion, students will realise that the nervous system controls the heart and lungs, so if the brain or nerves stop working, the heart and lungs will die. Also, if the lymphatic system or renal system stops working, the backup of waste products will cause toxic effects, ultimately resulting in death. So ... which is the most important organ? Which is the most important system?

Refractory Period – Toilet Flushing

Again, this uses a striking mental image, which is bordering on toilet humour and will be very memorable.

Neuron firing can be likened to the flushing of a toilet. A toilet is flushed by introducing more water, which occurs by pushing a lever. The water remains in the cistern until the threshold is met and the lever is activated, at which time it all falls through. The changes in water level in the cistern are analogous to voltage changes in the neuron caused by synaptic activity.

So, the resting potential would be what happens when the toilet has finished flushing; the action potential would be represented by water coming in when the lever is pressed, but a toilet never goes above threshold since there is a ballcock preventing overfilling. The refractory period is the time after flushing before the toilet can flush again, while the cistern is filling. The flow is in one direction only – in the case of water, with gravity.

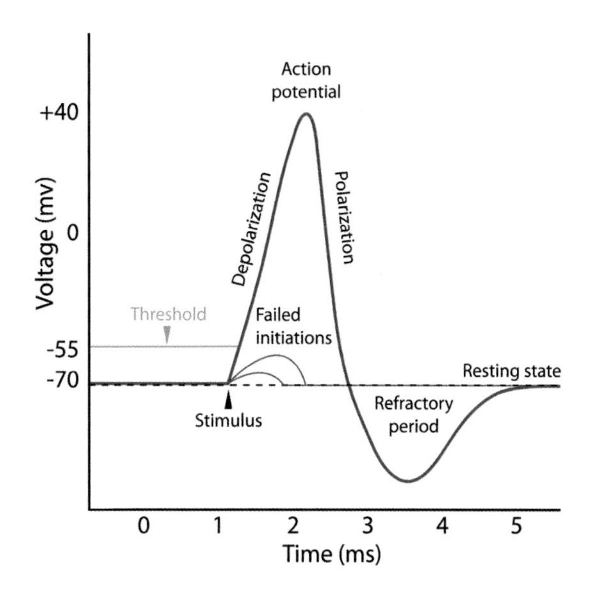

Figure 9.16: The refilling of a toilet cistern is analogous to the refractory period.

Role Play

Not everyone enjoys role play, so the mood and needs of the class should be considered before role play is used. If role play is to be used as a learning tool, it simply must be fun in order to be memorable. If the class is not in the right mood, they will only remember how horrified they were to act in a daft play, rather than remembering the key facts. Pitched to the right audience and delivered in a light-hearted and humorous way, role play can be a really effective learning tool for anatomy and physiology. Back up the discussions by creating a storyboard for the role play.

Do remember that some people learn by doing, others by watching. Make sure everyone gets a chance to take part in each position and is also in the audience, but avoid going through the 'performance' too many times, to prevent overkill and boredom. Keep it fresh!

Body-Part Charades

A good, old-fashioned parlour game of body-part charades is great fun. Giving a student an anatomical structure to 'act out' is thought-provoking for both the actor and the audience. They must consider the size and shape of the structure, as well as its functions. This can also be done as group work, so that half the class collectively acts out the functions of a structure and the other half of the class attempts to guess its name.

Human Sculpture

Another active learning exercise is modelling human sculpture. This requires full consent by all participants as it requires the students to physically touch their fellow students in order to move their bodies into shapes, matching movements to magazine pictures using anatomical language.

Students are divided into groups. One person is designated the 'lump of clay' who is to be moulded into shape, one person is the 'sculptor' and the rest are the directors. The directors choose a picture from a magazine of a person who is standing, sitting or lying in a particular way. Using anatomical language, they direct the sculptor to shape the lump of clay until he or she resembles the chosen image. Terms such as *supine*, *prone*, *inferior*, *proximal* and *lateral* can be applied. Human sculpture is hugely fun, and groups enjoy the challenge. The level of learning can be increased by using the names and actions of muscles – for instance, an image of a man kicking a football could be sculpted by asking the lump of clay to flex the right psoas whilst dorsiflexing by contracting the anterior tibialis.

Immune-System Role Play

Role play is used to show how the immune response acts to destroy a particular pathogen (bacterium or virus). Each student is given a specific role in the immune system (antibody, helper T cell, cytotoxic T cell, B cell, complement, etc.), and then the group discusses how each of them is involved in the destruction of the pathogen. Eventually they act out the 'war on bugs'. The students should complete a storyboard to record the processes involved in pathogen destruction.

Spinal Role Play

Acting out a structure as a group can be a fun and memorable way of learning its component parts. For instance, lining up the class to form a spine.

Instructions:

1. Line up the class in single file.
2. Put a large inflated balloon between each person, level with the belly, so that it becomes an intervertebral disc.
3. Have students flex their elbows to bring the forearms parallel to the floor.
4. Place broomsticks in the crook of their elbows, passing behind their backs, forming transverse processes.
5. Make the students close up, so their hands and elbows form 'facets' joining up with the person in front and the person behind.
6. Students' bodies become the vertebral bodies.
7. Their heads form the spinous processes (and can be further enhanced by wearing cone-shaped party hats!).
8. 'Bend' the spine by moving the class, to demonstrate how much the transverse processes move in any rotation.
9. Compression of the spine by having the 'vertebrae' move closer together will squash the balloons, or intervertebral discs, rather than squash the vertebral bodies.

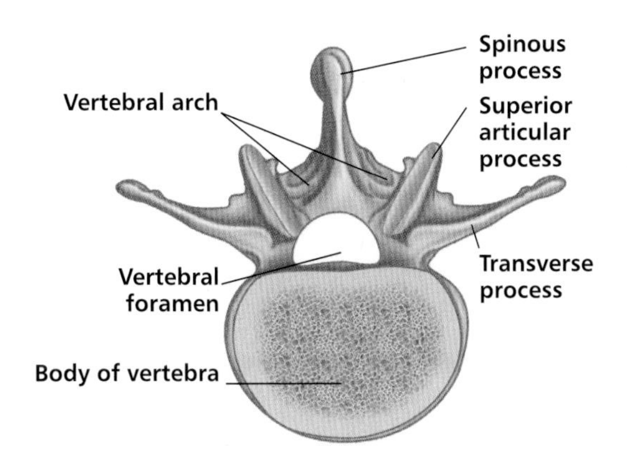

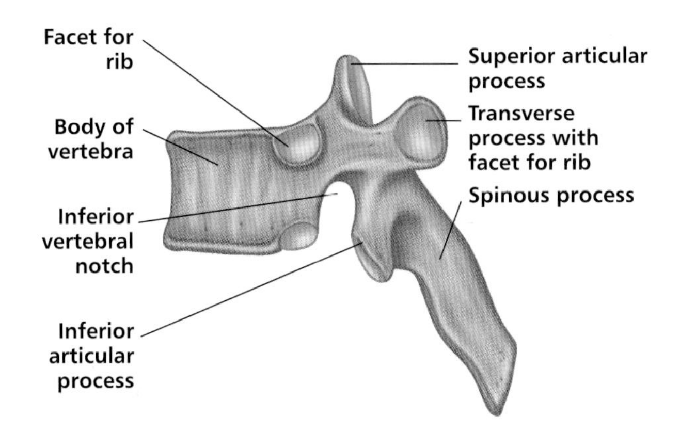

Figure 9.17: Diagram of vertebra showing transverse processes and facets.

Neurology Role Play

Neurology lends itself to role play as it involves pathways of neurons, which communicate with each other. Additionally, there is an exchange of ions across the membrane, which can be acted out too.

Action Potential

Students form a circle to represent one excitatory cell (cardiac pacemaker or myocardial cell). Holding different-coloured balloons representing ions (red – Na^+, blue – K^+, green – Ca^{++}), students move the ions in and out of the circle/cell to demonstrate the depolarisation/repolarisation state of the cell.

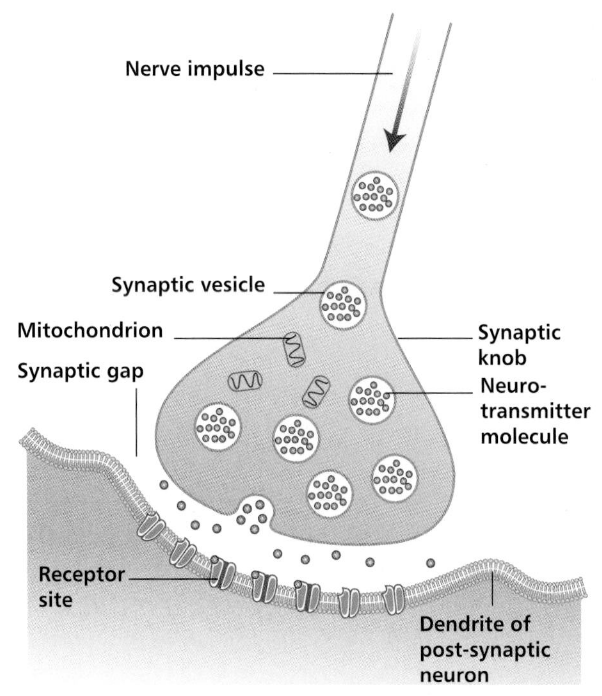

Figure 9.18: Illustration of neuron and action potential.

Similar role play could be used to demonstrate the cardiac cycle or the action potential proceeding between nodes of Ranvier, with each person playing the part of a synapse or node of Ranvier. The signal could be passed as a hand clap with the neighbouring student, or an exchange of 'neurotransmitters' in the form of tennis balls across a gap.

Speed-Dating-Style Game, Describing the 'Personality' of a Muscle or Structure

This game is loosely based on the premise of 'speed dating'. The game involves a student taking on a 'muscle' personality when introducing him- or herself to potential mates. Students assume the personality of the muscle that is listed on the name card attached to a nametag and are given two minutes to introduce themselves to a succession of possible antagonists and agonists, giving their origin, insertion, actions, planes of motion, and how they relate to functional activities as their chat-up lines. After the speed dating is over, the students can mingle and pair themselves up with perfect match of antagonist or agonist.

The students are removed from the traditional classroom lecture and are able to participate in an activity that is related to a social activity. The activity is fun and provides visual, physical and verbal associations that students may remember when recall is necessary. The anatomy information is not only taught by the class instructor but is relayed by fellow students. The students play an active role instead of a passive role, as is often seen in a kinesiology and functional anatomy classroom (McCarroll et al., 2009)

Anthropomorphism

When we describe our much-loved dog as a 'faithful old girl' or say that our old, clapped-out car is 'on its last legs' we are usually unaware that we are using anthropomorphism to give human characteristics to an animal or object. We use such language as it is descriptive and familiar, and gives emotive behaviour to otherwise inanimate objects or nonhuman species. Such description is memorable and often funny, especially when used for body parts.

If You Were a Body Part, What Body Part Would That Be and Why?

This is a great conversational game to use either in class or as a study session with a study buddy. Ask yourself and each other which body part you would be and why. It can be tailored to a particular body system. Using the functions and structures of body parts, you can identify yourself with an anatomical structure. For instance, thinking of cranial bones, each bone could have a different personality based on its shape and location. For instance, one of the students may be a somewhat flighty soul, and he or she could be likened to the sphenoid, as the beautiful structure of its wings can be likened to a butterfly, bird or bat. Someone who is 'straight down the middle' and direct could be the vomer, with its sword-like shape, in the middle of the head.

Thinking of the digestive system, a 'mouth' person could be someone who likes to chew things over, and may swallow his or her words, so in fact be rather quiet. The duodenum-type student likes to digest things and gets a lot of input from others, compared with the ileum of the group who absorbs information like a sponge. The class colon has a dry sense of humour and likes to consolidate things and deal with information in bulk, but tends to be full of hot air. The less said about the rectum of the class, the better!

Give anatomical structures personalities to help memorise functions. Thinking of hormones, it can be seen that the stressed hyperactive student would be adrenaline, pushing into flight and fight, whereas the student who is rather tall could be growth hormone. Once you start doing this it becomes quite fun, and all sorts of personalities can be attributed to the organs and body parts, and the fabulous thing is that these will be memorable forever. Obviously, students should be sensitive to the feelings of others, and not make comparisons that may be hurtful or insulting.

Games

Children learn through play; yet, when we begin to learn an academic subject, often the fun disappears. In practice, most adults will enjoy some fun and games in class, particularly if the level of learning is maintained and the information is not too 'dumbed down'. These games can be adapted to suit all levels of students and are great fun.

Who Am I?

This is an old parlour guessing game using headbands or a hat with a ribbon or band on it. Students take it in turn to wear the headband around their forehead. Cards with the names of diseases, anatomical structures or physiological functions are placed face down on a table, and one is chosen and placed in the headband so that the wearer is unaware of its name. The wearer has to ask questions of the rest of the audience that have either a 'yes' or a 'no' answer.

For instance, suppose the selection of cards was names of bony landmarks. Possible questions for cards requiring the naming and sequencing of bony landmarks for guessing the coracoid process could be:

- Is it on a long bone?
- Is it above the waist?
- Is it on both sides of the body?
- Do I have two of them?
- Is it a muscle attachment?
- Is it the attachment for some chest muscles?
- Is it the attachment for an arm muscle?
- Is at the origin of the pectoralis minor?
- Is it at the origin of the coracobrachialis muscle?
- Is the coracoid process?

Snap or Happy Families

Flash cards can be used to find similar structures based on similar location, or attachments. For instance, all the muscles which are innervated by the spinal nerve at C5 to C8 could become one family, those innervated by C1 to C4 could be another family.

Likewise, muscles which share an origin may be a family, and considering the interlinking of muscles needed to maintain structure will be visually and academically memorable. Looking for interconnections and interrelationships is hugely valuable as they forge links in the brain, and become committed to the long-term memory as it begins to 'make sense', rather than being a list of things to be learned by rote.

Jigsaws

Some people enjoy doing jigsaws. They enjoy the challenge of looking at the big picture, and finding the right shape or the right colour to place the right piece into the puzzle. Although anatomical jigsaws are hard to find, they can be made from taking a large print of an anatomical picture and mounting it on card or hardboard, then cutting up into shapes.

Jigsaws made of only 8–12 pieces can be used as a prize in quizzes. The aim of the game would be to put together all the pieces of a jigsaw puzzle, by answering correctly as many questions as possible. Ideally the anatomical picture should related to the questions that are being asked.

This is a good way of embedding information, as there is a visual reminder of what is being asked about, which is thus memorised. When sitting in an exam, recalling the game and the picture on the puzzle may trigger details to be yanked from the memory banks.

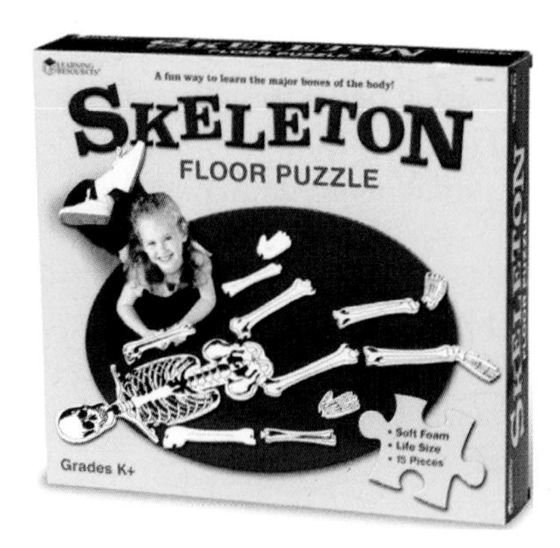

Figure 9.19: A skeleton floor puzzle. From educationaltoysplanet.com.

Board Games

Anatomix is a manufactured board game from The Green Board Game Company that helps you learn about your body and have fun at the same time. Although this is really designed for children, it can be adapted for use with adult learners by writing some more questions, or demanding more detail in the answers. The winner is the first to build their body using cards representing the body systems. You can choose between nerves, skeleton, organs and muscles; body pieces may have to be swapped or donated to the body bank, just like in real life! The game is cleverly designed to enable adults and youngsters to easily play together combining different levels in one game. The advanced player may have to wiggle their gluteus maximus or guess whether they can swallow when they stand on their head. Anatomix educates and entertains – with laughs and learning along the way.

Figure 9.20: Anatomix is a good game to help you learn about your body.

Twister – for Movements and Locations

Manufactured by Hasbro, the game Twister has been around for years. The premise can be converted for anatomical use by substituting the spinner for anatomical locations or actions that are selected by picking a card from a pile. For instance, choosing *flexion*, *left*, *knee*, means that the student has to flex his or her left knee and maintain position until their next turn, where the next choice of card might mean extending the right arm, and thus having to stand with left knee bent and right arm out. The winner is the one who remains standing the longest! Clearly, consent must be given by participants, in that students will be touching each other.

Songs, Poetry and Mnemonics

Learning (and memorising) the names and locations of anatomical structures isn't easy, so clinical anatomy students often develop mnemonics, or memory tricks, to make it a little easier. These mnemonics include acronyms, short poems, and silly phrases that are quite effective for remembering parts of the body. The dafter and sillier (and, often, ruder!), the better it is to help you remember. Borrowing other people's mnemonics is never as good as making up your own, so don't just rely on these, get inventing!

The Cranial Bones

Your skull has six cranial bones that form the cranial vault. You don't want to confuse them with the facial bones, so you can remember them with this phrase:

PEST OF 6

Each letter stands for a cranial bone, and the number 6 reminds you that there are six of them:

- P: parietal bone
- E: ethmoid bone
- S: sphenoid bone
- T: temporal bone
- O: occipital bone
- F: frontal bone

The Facial Bones

Your face is formed by eight facial bones. Here's a silly saying to help you remember them:

Virgil can not make my pet zebra laugh

- V: vomer
- C: conchae (inferior)
- N: nasal bones
- M: maxilla
- M: mandible
- P: palatine bones
- Z: zygomatic bones
- L: lacrimal bones

The Cranial Nerves

The cranial nerves each have a name and a number, so remembering which name goes with what number can be difficult. Use this little poem to remember:

On old Olympus's towering top, a Finn and German viewed some hops

- O: cranial nerve I, olfactory nerve
- O: cranial nerve II, optic nerve
- O: cranial nerve III, oculomotor
- T: cranial nerve IV, trochlear
- T: cranial nerve V, trigeminal nerve
- A: cranial nerve VI, abducens nerve
- F: cranial nerve VII, facial nerve
- A: cranial nerve VIII, auditory (or vestibulocochlear) nerve
- G: cranial nerve IX, glossopharyngeal nerve

- V: cranial nerve X, vagus nerve
- S: cranial nerve XI, spinal accessory nerve
- H: cranial nerve XII, hypoglossal

The Heart-Valve Sequence

The following odd sentence helps you remember how blood flows through the heart, by remembering the sequence of the valves:

TRy PULling My AORTA

The capital letters represent the valves, in order of blood flow:

- TR: tricuspid valve
- PUL: pulmonary valve
- M: mitral valve
- AORTA: aortic valve

Ordering the Abdominal Muscles

One way of remembering the names of the abdominal muscles is to think of a spare tire (we have to use the American spelling here, or it won't work!), which is the nickname for the extra fat that can build up around a person's abdomen.

The word *TIRE* stands for the four abdominal muscles:

- T: transversus abdominis
- I: internal abdominal oblique
- R: rectus abdominis
- E: external abdominal oblique

The Intestinal Tract

The intestinal tract includes the small intestine, the colon and the rectum. Use this phrase for remembering the parts of the intestinal tract and their sequence:

Dow Jones industrial climbing average closing stock report

The first letters of the first three words represent the three segments of the small intestine. The rest help you remember the colon and the rectum:

- D: duodenum
- J: jejunum
- I: ileum
- C: caecum
- A: appendix
- C: colon
- S: sigmoid colon
- R: rectum

The Rotator Cuff Muscles

Four rotator cuff muscles run from the scapula to the humerus and work together so that you can rotate your arm. They're usually remembered by the acronym *SITS*:

- S: supraspinatus
- I: infraspinatus
- T: teres minor
- S: subscapularis

The Carpal Bones

Eight carpal bones form the wrist. They're arranged in two rows, with four bones in each row. Following is a phrase that can help you remember them:

She looks too pretty; try to catch her

or

Some lovers try positions that they can't handle!

The first letters stand for each carpal bone. The first four form the proximal row (closer to the arm) starting laterally (thumb side). The second group of four make up the distal row (closer to the hand), also starting laterally:

- S: scaphoid
- L: lunate
- T: triquetrum
- P: pisiform
- T: trapezium
- T: trapezoid
- C: capitate
- H: hamate

The Lateral Rotator Muscles of the Hip

Six hip muscles rotate the hip laterally. Following is a phrase that will help you remember them:

Piece goods often go on quilts

The first letters represent the hip rotators, in order, from most proximal to most distal:

- P: piriformis
- G: gemellus superior
- O: obturator internus
- G: gemellus inferior
- O: obturator externus
- Q: quadratus femoris

The Tarsal Bones

You can remember the tarsal bones with this sentence:

The circus needs more interesting little clowns

- T: talus
- C: calcaneus
- N: navicular
- M: medial cuneiform
- I: intermediate cuneiform
- L: lateral cuneiform
- C: cuboid

Songs and Music

If you are musically talented, or just like a good sing-song, then perhaps putting your anatomical knowledge into song and verse could help you to remember key facts.

12 Cranial Nerves:

www.youtube.com/watch?v=_RJLLiP2hYo

Bones and their functions:

www.youtube.com/watch?v=EfWzcXmpUkc

http://seasonsstudios.com/anatomix-comix

Revision Walks

This is a very different type of revision and study tool which you may not have thought about doing. Do you like to go for a walk? Are you a dog walker? Then use this time to listen to audio recordings of some of your textbooks. If you walk the same route regularly, and listen to the same recording each time, you will associate different parts of your route with different facts that you have been listening to. By replaying your route in your mind, you will be able to recall it. Try choosing a route for each body system, or for each topic you are studying.

Even if you don't listen to audio recordings, you can walk and 'label' certain items on a well-known route as anatomical structures. Spend some time at each of these places, thinking and perhaps reading about a topic. When you think of the route in your mind's eye, you will be able to visualise the anatomical structure easily.

Anatomy Toys

Science Museum Fluffy Body Parts

When you attend science museums, take some time to mooch about in the shop. A plethora of postcards, models, cuddly toys and gifts can often be purchased and these can be used to demonstrate in class, or to remind a student of a key fact. Cuddly neurons and antibodies have their place in class!

Jacob's Ladders for Cruciate Ligaments

Attempting to imagine the orientation and function of the cruciate ligaments is often difficult for students. Take a traditional wooden toy, the Jacob's ladder, which has the same layout in its cross-threading, and show how this layout of 'ligaments' gives absolute stability in one plane of movement, therefore needing collateral ligaments supporting and preventing excessive sideways movement.

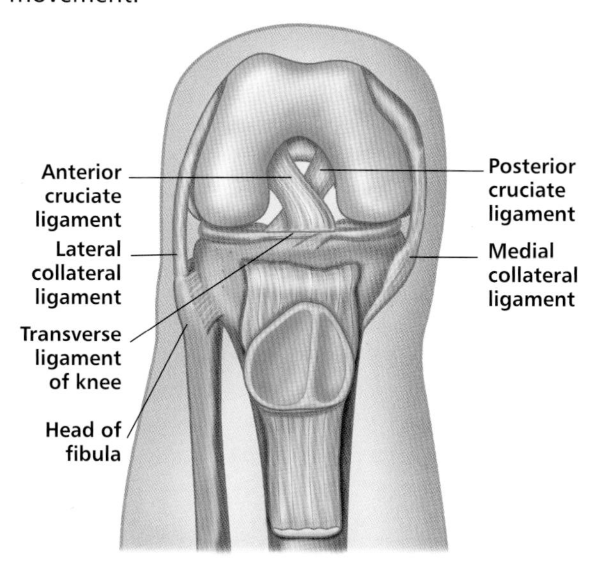

Figure 9.21: The knee ligaments are like a Jacob's ladder.

Children's Anatomy Kits for Heart, Brain and Digestive System

These can be purchased to build plastic models of anatomical structures. Perhaps not as much fun as the 'home spun' models, but usually more accurate in proportions and detail. Amazon offers a huge range of anatomical toys. Educational Toys Planet is an online toy shop based in New Jersey, USA. It sells a range of toys; in particular it sells anatomical toys, games and puzzles for children of all ages (educationaltoysplanet.com).

Anatomy Gifts

CafePress (cafepress.co.uk) has thousands of anatomically related gifts, such as pillow cases, mouse mats, mugs, vinyl stickers, shower curtains and blankets printed with diagrams, artwork, X-rays, photographs and photomicrographs; jewellery made from tiny photomicrographs; and even a pair of skeleton flip flops! These are great as prizes for revision classes, rewards for great homework or just a purchase to help the revision.

Clothing is also available from Seasons Studios (seasonsstudios.com). They also make anatomical comics called Anatomix Comix, designed for children as well as for adults with a sense of fun (amazon.co.uk).

Useful Apps

Technology is incredibly fast moving, and new apps for devices are being produced daily. The use of mobile devices gives great opportunities for learning and revision on the move and without the need for textbooks or scraps of paper.

Current excellent apps include:

- Essential Anatomy 5 (3d4medical.com)
- Visible Body (visiblebody.com)
- Netter's Anatomy Flashcards (modality.com)
- AnatomyMapp (booksofdiscovery.com)

Useful Websites

Current websites that should be bookmarked for anatomical help are:

booksofdiscovery.com

Books of Discovery website is to support students and teachers who used the *Trail Guide to the Body* textbooks. This palpatory guide to anatomical structure and landmarks is an invaluable tool for students of manual therapies, and is available in different formats: printed book, DVD, audio and PowerPoint – so fulfils most students' learning styles.

For teachers and instructors, the website offers lesson plans, PowerPoint presentations and images that can used for lectures, and also offers the 'Instructors Back Pack', which gives hints and tips for creative lessons.

anatomyzone.com

Anatomy Zone is a free website that offers a comprehensive range of three-dimensional (3D) anatomy tutorials, covering all topics of anatomy, delivered in a relaxed manner. The 3D images give the visual learner a full appreciation of location of anatomical structures, and it is narrated in a very relaxed way, so you meander through the human body, taking in the awesome sights as you go.

eorthopod.com

eOrthopod also offers narrated video tutorials for students and teachers, as well as for the general public, aiming to inform and educate about anatomical structures. It also includes orthopaedic fact sheets about certain diseases and injuries, which are in clear and simple, yet anatomically correct, language, making it understandable to the public, without overly 'dumbing down'.

kenhub.com

Ken Hub offers video tutorials, articles and quizzes to support anatomy and physiology lectures. Some articles and videos can be accessed free of charge, but, by paying a monthly subscription, more articles and videos are made available.

innerbody.com

A free resource with hundreds of interactive anatomy pictures and descriptions of thousands of objects in the body, InnerBody will help you discover what you want to know about human anatomy, in a point-and-click manner. It is simple in design, yet fairly detailed.

primalpictures.com

Primal Pictures' 3D anatomy model, built using real scan data from the Visible Human Project, has been carefully created with an unparalleled level of detail and accuracy. All of the content has been verified by qualified anatomists and by a team of external experts for each body area. The modular interactive 3D anatomy software allows you to engage with 3D models, using intuitive functions to rotate, add or remove anatomy, and identify and learn more about any visible structure. Case studies give real medical context and relevance to the learning process, while revision aids, such as the TEST function and multiple-choice questions (true/false/pass), help you learn what you need to know. Although it is pricey, it may be worth its weight in gold if you enjoy learning in this way.

nhs.uk

The NHS website offers a wealth of informative articles and videos, for both patients and professionals, about health and disease, which will support an anatomy and physiology course that goes into the pathologies of the anatomy.

3d4medical.com

3D4Medical is an award-winning technology company that specialises in the development of medical, educational, and health and fitness apps for professional reference as well as student and patient education. Their apps are expertly designed to encourage unique learning experiences through intuitive interactivity and stunning visualisations.

3D4Medical's latest ground-breaking apps have revolutionised the teaching and learning of anatomy, physiology and exercise, enabling users to effortlessly navigate via 360° views of the human body. Their apps have also proven to be of enormous benefit to healthcare professionals as a means by which to illustrate and communicate effectively with patients, pupils and clients. The highly detailed interactive visualisations, combined with comprehensive and searchable indexing systems and quiz functions, also make the apps indispensable learning resources for students.

studyblue.com

Study Blue is an online tool for building flash cards. These can be made by students, either for their own use, or to be used by schools who wish to monitor their students' progress. Study Blue is a collaborative learning ecosystem that is used by more than 10 million people. Students can connect with others who are on a similar learning journey through a shared library of more than 250 million pieces of user-generated content. By selecting the topic 'Anatomy and Physiology', access to a vast library of flash cards and resources is given.

getkahoot.com

Kahoot!'s free game-based platform engages the heart, hand and mind to create a more social, meaningful and powerful pedagogical experience. With Kahoot!, you can create, play and share fun learning quizzes in minutes – for any subject, in any language, on any device, for all ages. There are many preconstructed anatomy and physiology quizzes on the site. Kahoots are best played in a group setting, like a classroom. Players answer on their own devices, while games are displayed on a shared screen to unite the lesson – creating a 'campfire moment' – encouraging players to look up.

Final Thoughts

Resources for anatomy are constantly changing, evolving and being developed. Learning and teaching involves the integration of resources and learning how to use them effectively. As 'smart' technology continues to develop, we as a community of teachers and learners begin to realise the endless possibilities of how the study of the human body is being transformed. Visit learnanatomy.uk for further updates and reviews of learning and teaching resources.

Reference

McCarroll, M.L., Pohle-Krauza, R.J. and Martin, J.L. (2009). Active learning in the classroom: A muscle identification game in a kinesiology course. *Advances in Physiology Education* **33(4)**: 319–322.

A Appendix

The VARK Questionnaire (Version 7.0)

How Do I Learn Best?

Choose the answer which best explains your preference and circle the letter(s) next to it. **Please circle more than one** if a single answer does not match your perception.

Leave blank any question that does not apply.

1. You are helping someone who wants to go to your airport, town centre or railway station. You would:
   a. go with her.
   b. tell her the directions.
   c. write down the directions.
   d. draw, or give her a map.

2. You are not sure whether a word should be spelled `dependent' or `dependant'. You would:
   a. see the words in your mind and choose by the way they look.
   b. think about how each word sounds and choose one.
   c. find it in a dictionary.
   d. write both words on paper and choose one.

3. You are planning a holiday for a group. You want some feedback from them about the plan. You would:
   a. describe some of the highlights.
   b. use a map or website to show them the places.
   c. give them a copy of the printed itinerary.
   d. phone, text or email them.

4. You are going to cook something as a special treat for your family. You would:
   a. cook something you know without the need for instructions.
   b. ask friends for suggestions.
   c. look through the cookbook for ideas from the pictures.
   d. use a cookbook where you know there is a good recipe.

5. A group of tourists want to learn about the parks or wildlife reserves in your area. You would:
   a. talk about, or arrange a talk for them about parks or wildlife reserves.
   b. show them internet pictures, photographs or picture books.
   c. take them to a park or wildlife reserve and walk with them.
   d. give them a book or pamphlets about the parks or wildlife reserves.

6. You are about to purchase a digital camera or mobile phone. Other than price, what would most influence your decision?
   a. Trying or testing it.
   b. Reading the details about its features.
   c. It is a modern design and looks good.
   d. The salesperson telling me about its features.

7. Remember a time when you learned how to do something new. Try to avoid choosing a physical skill, eg. riding a bike. You learned best by:
   a. watching a demonstration.
   b. listening to somebody explaining it and asking questions.
   c. diagrams and charts – visual clues.
   d. written instructions – e.g. a manual or textbook.

8. You have a problem with your knee. You would prefer that the doctor:
   a. gave you a web address or something to read about it.
   b. used a plastic model of a knee to show what was wrong.
   c. described what was wrong.
   d. showed you a diagram of what was wrong.

9. You want to learn a new program, skill or game on a computer. You would:
   a. read the written instructions that came with the program.
   b. talk with people who know about the program.
   c. use the controls or keyboard.
   d. follow the diagrams in the book that came with it.

10. I like websites that have:
   a. things I can click on, shift or try.
   b. interesting design and visual features.
   c. interesting written descriptions, lists and explanations.
   d. audio channels where I can hear music, radio programs or interviews.

11. Other than price, what would most influence your decision to buy a new non-fiction book?
   a. The way it looks is appealing.
   b. Quickly reading parts of it.
   c. A friend talks about it and recommends it.
   d. It has real-life stories, experiences and examples.

12. You are using a book, CD or website to learn how to take photos with your new digital camera. You would like to have:
   a. a chance to ask questions and talk about the camera and its features.
   b. clear written instructions with lists and bullet points about what to do.
   c. diagrams showing the camera and what each part does.
   d. many examples of good and poor photos and how to improve them.

13. Do you prefer a teacher or a presenter who uses:
   a. demonstrations, models or practical sessions.
   b. question and answer, talk, group discussion, or guest speakers.
   c. handouts, books, or readings.
   d. diagrams, charts or graphs.

14. You have finished a competition or test and would like some feedback. You would like to have feedback:
   a. using examples from what you have done.
   b. using a written description of your results.
   c. from somebody who talks it through with you.
   d. using graphs showing what you had achieved.

15. You are going to choose food at a restaurant or cafe. You would:
   a. choose something that you have had there before.
   b. listen to the waiter or ask friends to recommend choices.
   c. choose from the descriptions in the menu.
   d. look at what others are eating or look at pictures of each dish.

16. You have to make an important speech at a conference or special occasion. You would:
   a. make diagrams or get graphs to help explain things.
   b. write a few key words and practice saying your speech over and over.
   c. write out your speech and learn from reading it over several times.
   d. gather many examples and stories to make the talk real and practical.

The VARK Questionnaire Scoring Chart
Use the following scoring chart to find the VARK category that each of your answers corresponds to.

Circle the letters that correspond to your answers, e.g. If you answered b and c for question 3, circle V and R in the question 3 row.

Question	A	B	C	D
3	K	(V)	(R)	A

Scoring Chart

Question	A	B	C	D
1	K	A	R	V
2	V	A	R	K
3	K	V	R	A
4	K	A	V	R
5	A	V	K	R
6	K	R	V	A
7	K	A	V	R
8	R	K	A	V
9	R	A	K	V
10	K	V	R	A
11	V	R	A	K
12	A	R	V	K
13	K	A	R	V
14	K	R	A	V
15	K	A	R	V
16	V	A	R	K

Calculating your scores
Count the number of each of the VARK letters you have circled to get your score for each VARK category.

Total number of **V**s circled =
Total number of **A**s circled =
Total number of **R**s circled =
Total number of **K**s circled =

Calculating your preferences
Use the VARK spreadsheet (which can be purchased from the www.vark-learn.com web site) to work out your VARK learning preferences.

Index